ISLAND

DIARY OF A YEAR ON EASDALE

Vicky and Garth Waite

MAINSTREAM PUBLISHING

EDINBURGH

CENTURY PUBLISHING

LONDON

For

   Christopher,
   Glenda,
     and
  all our friends
  North and South
   of the Border

~ HEARTSEASE ~
from the garden

Pansies are for thoughts

First published in Great Britain in 1985
by Century Publishing Co. Ltd,
  Portland House
12-13 Greek Street, London. W1V 5LE
in conjunction with
Mainstream Publishing
 7 Albany Street,
 Edinburgh. EH1 3UG

British Library Cataloguing in Publication Data

Waite, Garth
  Island: diary of a year on Easdale
  1. Easdale Island (Scotland) Social Life and Customs
  1. Title II. Waite, Vicky
  941. 4'23 DA880.E/

ISBN 0 7126 0766 8

Printed in Great Britain
by Purnell & Sons (Book Production) Ltd, Paulton, Bristol

~CONTENTS~

Introduction	iv - xv
Map	xvi
Diary	1 - 140
Index	141 - 144

~ Introduction ~

In my possession is a first school report written at the end of Summer Term 1922 when I was five years old. Against the subject 'Nature Work' are the words 'Shows keen interest in everything'. The Cheltenham Grammar School kindergarten mistress deserved full marks for perception, for this interest has persisted all my life.

As a child I, my long-suffering baby sister and a few companions, spent most of our playtime out-of-doors on a farm where the last working cider-press was still operating. It was powered by a horse pulling a stone wheel over the apples in a circular trough. The fields were horse-ploughed and cultivated. Chickens, ducks, pigs and sheep occupied adjoining paddocks and occasionally broke through into the garden. The annual visit of the threshing-machine which was driven by a portable petrol engine on a crude chassis was an exciting event. As the corn stack was dismantled the mice ran in hundreds in all directions and small boys, armed with sticks, decimated their ranks.

SONG THRUSH

Systematically I collected live creatures - beetles, snails, newts and spiders - and also succeeded in breeding three-spined sticklebacks in an aquarium A cousin, with whom I grew up and in whose company I enjoyed many childish expeditions, became - and still is - a successful breeder of caged and aviary birds.

All this, and much more, was the repository from which I gathered, almost imperceptibly, a first-hand knowledge of things about which field naturalists and biologists have written. When, as a teenager, I acquired a second-hand and much annotated copy of the awkwardly titled 'Observations in Natural History with an introduction on Habits of Observing: Also a Calendar of Periodic Phenomena in Natural History', my mind was made up. The work of one Rev. Leonard Jenyns of Swaffham Bulbeck, Cambs, and published in 1846, this volume so enthralled me that I resolved to follow the author's example, not by entering the Ministry (though I am an accredited Preacher of the Methodist Church and a certificated Reader and Elder of the Church of Scotland) but by nourishing my innate keen interest and sharpening and refining my observations. This was the fertile soil in which the ambition to complete a nature notebook was grounded. It must be said, however, that the planting of the seed predated the fruit of the endeavour by more than fifty years, a span of seasons in which I lived successively in Gloucestershire, Middlesex, Lincolnshire and Hertfordshire, before retiring to Argyll, Scotland.

My first wife, Cecily, who bore me a son, Christopher, in 1951, died of disseminated sclerosis in 1974 after ten years in hospital. The years leading up to my retirement were spent working for the Methodist Church, where I shared in the administration at headquarters, Central Hall, Westminster.

It was while I was there that the word 'coincidence' became a reality for me and for someone else. The course of both our lives was to be changed dramatically.

SMALL COPPER BUTTERFLY

Vicky was born in 1919, breathing the air of Buckinghamshire. Her artistic talent was noted early and she was offered a scholarship at the Brighton School of Art. She never took it up, nor did she have any formal art training whatsoever.

She also had the gift of musical ability and at twenty-one became a freelance pianist, playing in open-air concerts. At around this time she met Jack, a serving air-crew member of the RAF, whom she married. Within two years Jack was killed when his plane crashed on take-off from Gibraltar.

After three years in the Women's Land Army in Oxfordshire where she looked after ducks, geese and turkeys and worked in the dairy, she felt the loneliness of isolation as well as the physical strain. Returning home to her ageing parents, she sought to help those with greater problems than her own by joining the Red Cross and nursing war-blinded women at St. Dunstan's. She kept up her piano practice and was accepted as a mature student by the Guildhall School of Music and Drama, where she gained her diploma as a teacher of pianoforte in 1962. Around this time, on impulse, she sent three paintings to the Royal Watercolour Society, and another to the Royal Institute of Painters in London. To her great surprise they were all exhibited.

A casual meeting with someone in Brighton was to change Vicky's life. This stranger had come from an island in Argyll called Seil. Her description of the tranquillity of that place painted a picture of the kind of solace Vicky was seeking and she decided to spend a holiday there. The delights of that un-polluted corner impressed her with a haunting feel-ing that some time she would return.

It was in the summer of 1975, after losing her parents, that Vicky decided to repeat her visit and flew to Glasgow. The train she joined at Queen Street station was bound for Oban. She entered a carriage of the open plan variety and, much to the astonishment of the one

HAREBELL
(SCOTTISH BLUEBELL)

passenger in it, marched the full length of the car-
riage to sit opposite him. His name was Garth Waite.
He was on his way, alone, to Lochboisdale on South Uist.
He had enjoyed a succession of holidays on Eigg, Rhum,
Shetland and Orkney, and the delights of the Scottish
islands had got under his skin.

Both of us were ready - nay, eager - for adventure. We
talked on the train of islands and natural history, of music
and religion. Names and addresses were exchanged before
we reached Oban that June evening in 1975. There we
parted - she for Seil and I for South Uist. From that
time until the winter of 1976 we did not meet again, but
we corresponded.

1976 was not a good year for Vicky. Things within her
that wanted to be expressed were bottled up. At a low ebb
of frustration she telephoned me and I agreed to visit her
at her flat in Hove. That was on 16 December. We had
both already made individual arrangements for Christmas which
we honoured. On New Year's Day, 1977, I went to Hove and bought
her (and Bertie, her budgie) back to my home in Welwyn, Hertfordshire,
to meet my father and my son. She was to have stayed two

Easdale
Argyll
Scotland

nights. I would not let her return and we were married on the 22 January 1977.

My daily commuting to Westminster continued until the summer, when we decided upon a delayed honeymoon. It was natural that we should choose to repeat our earlier train journey and stay, this time together, on Easdale Island, where we rented a cottage for a fortnight. One day I was fishing quietly in one of the flooded quarries when Vicky crept up behind me. 'What about retiring early and coming to live here?' she inquired. 'Give me an hour or two to think about it', I replied. The same day our decision was made. We mentioned the fact to Peter Long, the owner of the cottage we had rented. He and his wife, Margaret, had retired to Scotland from the south a year or two earlier. 'You've got to be a bit mad to come and live here', Peter said. It seemed to me that we were exceptionally qualified.

I gave in my notice to the Methodist Church and we bought a cottage that had become vacant. Our removal was accomplished in five overnight journeys in our own Bedford Work-o-Bus which we bought for the purpose. All our furniture and effects were transferred from the van to a little open ferry-boat. My son, Christopher, and several of his pals assisted us, and

Johnny McFadyen, the ferryman, always smiling, piloted us across the four hundred yards of swiftly-flowing current that separates Easdale island from Seil. It was August when we moved in. The mellow autumn was followed by the severest winter in living memory.

Our wee detached cottage has a floor area of 25 ft by 18 ft and is of the typical Scottish but-and-ben (two roomed house) type. Its modernisation divided the second room into two, thus providing a livingroom cum kitchen, a double bedroom (almost completely filled by the bed!) and a tiny bathroom. Electricity and mains water are connected: so is the telephone. More than two hundred years old the cottage, like its forty or so fellows, has walls more than 2 ft thick. It faces the rising sun and is fifty-five paces from the high-water mark. The building material – walls and roof – is slate.

From within the cottage it is easy for us to see what is happening outside – views are framed by windows back and front. The binoculars live on the table. From the front we can see the arrival of the 15 ft ferry-boat at the quay, always noting whether the mail is on its way. In the rough grass between the front door and the shore, finches often feed, and it is here that we have seen a snipe, a lapwing and a flock of snow buntings, besides the occasional strolling hen. The opposite window looks out at cushions of pink-flowered saxifrage, the slate-slab path and the rowan tree in the tiny garden. Wheatears and willow-warblers are the earliest spring visitors. Whitethroats also come, and a lone sedge-warbler. Several resident toads keep down the slug population, and shrews inhabit the compost heap. Our view through this west window takes in a collection of shells and other trophies gathered on the ledge, including the porcelain figure of the Chinese goddess of compassion Kuan Yin.

Our small library of natural history books, always at hand, enables us to verify every sighting. We also possess a Victorian volume, quaint but fascinating, on the meanings of flower-names. There is plenty to read when we sit round the old-fashioned fireplace on dark winter evenings.

The island itself, nowhere more than half a mile across, is composed of slate, and from it, over the centuries, thousands of tons have been quarried, fashioned, trimmed and shipped. The scars include seven water-filled quarries and many heaps of indestructible rubble.

The backbone of the island is a ridge, 105 ft high, running north-west to south-east. From the top are views of Garvellachs, Colonsay and Islay to the south-west, Mull to the north-west, and Lismore and Morvern to the north. Set in the Firth of Lorn, the island lies in the parish of Kilbrandon and Kilchattan. The footpaths (there are no roads) tend to follow the old tram-tracks where horse-drawn trucks of slates were hauled to the quay for export to Iona, Ireland and elsewhere. All this trade ceased in 1881, when a November gale swept a huge tide across the low-lying parts, flooding the quarries overnight.

A few Easdale cottages are still without electricity or mains water. One of them is next door to us. It is owned by Sadie and Margaret McCallum, a war-widow and her daughter, whose home is in Glasgow. Ancestral roots anchor them firmly to the island and they spend summer holidays and many weekends in their cherished corner. Sadie's brother, Peter Connelly, who used to manage the

CHAFFINCHES

post office with his wife across the water at
Ellenabeich and was once a ferryman, is a
large man with a large heart. He can put
his hand to anything - and usually does.
His almost fierce Scottish nationalism
is matched by a tenderness and sensitiv-
ity fed by his love of poetry which he
will recite with emotion. It was his - and
his family's - kind of warmheartedness that made incomers
such as ourselves grateful to be accepted into a tightly-knit
community where several cottages are still the property of the
proud descendants of islanders. Other cottages are now owned
by fishermen and builders as well as retired and active profession-
al folk including a doctor, an architect, a solicitor, a university
lecturer and a bank manager. The register of electors on the
island contains about thirty names. A handful of children cross
on the ferry daily to attend the village school.
    The only wild mammals on the island are water-shrews and
short-tailed voles. Toads proliferate and coarse grass, brambles
and blackthorn festoon the slate debris backed by vertical
rock faces which are adorned every spring with primroses and

violets. There are virtually no trees, and ground-nesting birds are persistently disturbed by resident dogs. The prevalent westerly winds bring much rain but the Gulf Stream tends to keep the worst frosts away.

Set in these idyllic surroundings our retirement afforded us the time and inspiration to bring about the marriage of our ambitions. Neither of us alone could accomplish what was attainable together. Vicky took her brush and pencil, pen and ink, and clothed my ideas with colour and script.

What is recorded here is the journal of our first splendid year on Easdale - 1979. It has taken us some six years to complete. In that time changes have occurred on the island: a new (absentee) owner, a new ferryman, several departures and a number of arrivals.

The slate heart of the island does not alter.

Garth Waite

St. Valentine's Day

1985

~ SCOTCH ARGUS BUTTERFLY
ON WHITE HEATHER ~

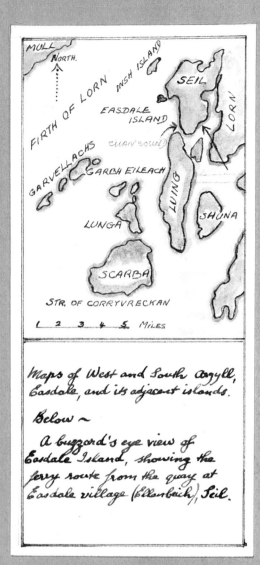

MULL

NORTH.

FIRTH OF LORN

INSH ISLAND

SEIL

EASDALE ISLAND

LORN

GARVELLACHS   CUAN SOUND

GARBH EILEACH

LUING

LUNGA

SHUNA

SCARBA

STR. OF CORRYVRECKAN

1 2 3 4 5 MILES

Maps of West and South Argyll, Easdale, and its adjacent islands.

Below ~

A buzzard's eye view of Easdale Island, showing the ferry route from the quay at Easdale village (Ellenabeich), Seil.

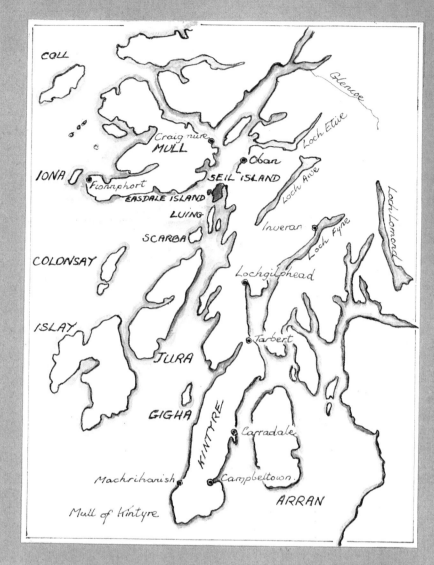

COLL

GLENCOE

Craignure

MULL

LOCH ETIVE

Oban

IONA

SEIL ISLAND

Loch Awe

Fionnphort

EASDALE ISLAND

LUING

Inveran

Loch Fyne

Loch Lomond

SCARBA

Lochgilphead

COLONSAY

Tarbert

ISLAY

JURA

GIGHA

KINTYRE

Carradale

Machrihanish

Campbeltown

Mull of Kintyre

ARRAN

JANUARY

~ COTTAGES
ON EASDALE
ISLAND ~

January

Ceud mhios na bliadhna ~ The first month of the year

WHIN.

January brings the snow
Makes our feet and fingers glow
Sarah Coleridge

Through the hushed air the whitening shower descends
At first thin wavering till at last the flakes
Fall broad and wide and fast
Dimming the day with a continual flow
Winter ~ James Thompson

The stormy north sends driving forth
The blinding sleet and snow
Winter ~ Robert Burns

The north wind doth blow
And we shall have snow
And what will the robin do then
~ poor thing ~

Jan 1
New Year's Day ~ Ne'er Day

The quietness of New Year's Eve with no breath of wind held a
surprise, for on waking we found the world outside all white with a
five~inch covering of snow. By sunrise a hungry robin and a cock
blackbird were at the door and we provided a mixture of soaked
bread and raw oat-meal on the window-sill. It was not long before
a second cock blackbird and a hen, as well as several hen chaffinches accepted the invitation to breakfast. With temperature near

A Robin
on
Angelica sylvestris ~
a prolific plant grow-
ing on the
island, which
blooms delicate
mauve and pink in
Summer, and seeds in late
Autumn.
It is related to the
Angelica, the stems of which
are crystallized and used
in confectionery.

freezing and snow still covering the ground, the number of birds coming to be fed increased. Several cock chaffinches joined their hens and soon we began to recognize individual birds, some by their particular markings ~ wider or narrower bands of white in the wings, and some by their habits in approaching the window.

Jan 2

A skein of Barnacle geese flying south passed overhead.
Country lore associates this behaviour with the coming of very cold weather in the near future. Through the binoculars we watched them alight on an uninhabited isle to the west.

Jan 3

The whiteness of the snow which lay round about was in strong contrast to the area of grey slate and brown seaweed alternately covered and revealed at high and low tide.

A grey heron, its feet in the icy water was patiently watching for a morsel to swim within range of its deadly spear-thrust.

It was here that small marine creatures came to the attention of a song-thrush, which, deprived of its more usual diet of land snails and worms hidden under the snow, quickly set up its characteristic hammering spot on a convenient stone. Inland thrushes do not have this opportunity.

PRIMROSE ~ Primula vulgaris
Meaning ~ Early youth

Her modest looks
        the cottage might adorn
Sweet as the primrose
        peeps beneath the thorn
            ~ Oliver Goldsmith ~

Cheerfully thou glinted forth
                amid the storm
Scarce reared above the parent earth
                thy tender form
                ~ Robert Burns ~

COMMON DAISY ~ Bellis perennis
Meaning ~ Innocence

Jan 4

A slight rise in temperature accompanied by a shower of rain has
started a gentle thaw. Tiny oases of green grow larger and larger
and the cushions of saxifrage in the garden look fresh and vigorous.
Here and there a brave primrose is in flower as if to salute the hardy
little daisies also blooming. Their presence under the snow had been
forgotten. I made the journey today into Oban. The going was not
easy as the road had not been gritted, but any difficulty in keeping
control on four wheels was exceeded by that of trying to walk on two
legs in the town where the pavements were clad in hard packed ice
and snow. I was glad to get back and impatient for the ferry boat to
meet me. While waiting I saw a mute swan near the quay. It was obviously hoping for something to eat and cocked its head on one side
to see me and my intentions more clearly. I walked along the quay

about five metres above the water level. Having just bought a sliced loaf I crumbled a couple of slices for the swan. It needed no persuasion and swallowed the pieces eagerly. Had it shown any delay or disinterest it would have lost its opportunity, as clamorous herring gulls were waiting near by.

Jan 5

Our community of birds
breakfasting from their slate
platter on the window-sill fled to-day
as a kestrel made a close approach before
veering away in search of something else. It was not long before they resumed feeding, among them a hen blackbird so tame that she habit-ually stays on the sill even while a fresh food supply is put down.

Jan 6

Recent high tides have deposited many kinds of seaweed at the high water mark, and on the shore a group of four hooded crows were peck-ing among the wrack; it seemed that their industry was very rewarding.

Exotic! ~ from the soil no tiller ploughs,
Save the rude serge ~
    fresh stripling from a grove
Above whose tops
    the wild sea-monsters rove.

~ An address to a piece
of seaweed ~

Schiller

~SERRATED WRACK~
Balanus balanoides
and
~ACORN BARNACLES~
Fucus serratus

Jan 7

Our afternoon walk took us to the west of the island where by 4 p.m. the sun was beginning to set. Its reddening rays from behind the gathering clouds lent a rosy pink glow to the distant island of Garbh Eileach, the largest of the Garvellachs. Garbh Eileach means Rough Island..

HOODIES on the west shore

KNOTTED WRACK.

Jan 8

After another fall of snow during the night we learnt of roads impassable because of drifts in many parts of Scotland and England.
Looking across the Firth of Lorn I watched four mute swans flying only about a metre above the water. Although so much of the ground is snow-covered, it is interesting to see how much snow-free ground is available to small birds under and around the gorse or whin as it is called here. Beneath these bushes is shelter and perhaps a little food for the greenfinches that are seen pecking about there.

Jan 12

There is special delight in seeing a species of bird observed for the very first time. This was our pleasure to-day when a flock of

REED BUNTING in company with SNOW BUNTINGS

nine or ten snow buntings were noticed not far from the cottage. With the binoculars, identification was certain, the orange bill, black legs, rusty smudge on the breast and white wings and underparts were clear to see. The birds seemed full of vitality as they pecked among the seeding rye grass and crested dog's tail protruding from the snow. A reed bunting in their company was a useful comparison. They were bigger than he, and from their general shape and behaviour conveyed for sure that they are not merely a different species, but another genus from the other buntings seen in Britain.

A party of skylarks were feeding near the buntings, and each time they flew up their white tail patches flashed clearly, and they uttered their familiar call note. A single pee-wit investigating the bay at low water flew off as I approached, shouting its name.

MAIDENHAIR SPLEENWORT
Asplenium trichomanes

SEA SPLEENWORT
Asplenium marinum.

Fern, means ~ Sympathy

Jan 13
Brushing away the snow from one of the many slate walls about the island, we found little plants nestling in the crevices. Trailing stonecrop, wisps of thyme, mosses, lichen and several attractive little ferns. All of them seemed to be flourishing regardless of the weather.

STONECROP

Jan 14
I have worked out by reference to the nautical almanac that the times of high and low tides in any one place are constant in relation to the occurrence of the moon's phase. At low tide passengers have to scramble down the rocky shore to reach the ferry.

Jan 15

On a walk to-day I saw a party of purple sandpipers on the spray-drenched rocks at the western end of the island. They were easy to identify because of their yellow legs and the yellow coloured base of their bills.

Jan 16

It is fascinating to watch a bird whose behaviour immediately suggests its name. I was standing quietly and looking into a rock pool when a bird flew down almost at my feet. Its outstanding features were the bright orange legs and rather dull plumage. Paying no attention to myself it immediately set about turning over the pebbles at the water's edge, gobbling up the creatures so suddenly exposed, of which there were many.

I was surprised that it succeeded in tackling fairly large stones by wedging its closed bill underneath and giving a sharp toss of the head. That such a bird is called the turnstone makes recognition relatively simple. These birds commonly feed in parties sometimes associating with other species. No doubt there is some advantage in being close to the freshly turned stone to seize the prize before the bird that moved it can make its grab.

Jan. 19

The dense though stunted
bushes of blackthorn bare
of leaves which grow abun-
dantly in the shelter of
the ridge are adorned
with parmelia lichen.
This clusters round
the branches and
provides a substi-
tute decoration
at a time when
there are no
leaves, blos-
som or
fruit.

Blackthorn
Twig

From Field Notebook

How variable the many
species of lichen are. In
many places one treads
upon the tiny cup lichen
~ Cladonia, which is often found
together with moss on the ground.

Jan. 22   THYME

Walking together on our second wedding anniversary
we came to a place where a trickle of fresh
water emerges from the vertical face of the
ridge. This would seem to be surface water
from the cap, and when temperatures are
as low as they are just now, great icicles
form like stalactites. There are few places
where fresh water stands for long because of
the many fissures in the slatey subsoil, but
one or two small ponds allow the breeding of
toads which have disappeared into hibernation. In
the Spring we shall be looking out for their annual
migration from their hiding places to fresh water.

Jan 25 ~ Robert Burns's birthday.
Scots all over the world salute this date with Burns Night celebrations.
Two fieldfares appeared near the cottage, very obviously of the thrush family in their shape and movements. During an afternoon snow-shower a cock blackbird came to feed at the window. A snow-flake lodged just above his eye and remained there looking like a false eyebrow as he pecked away at the crumbs we had provided.

GREY SEAL

Jan 31
Seals are not uncommon in these waters, but they are very wary and do not betray their presence readily. Characteristically one grey seal pushed her nose end out of the water as I watched at about five hundred metres. After about half a minute she dived, repeating the performance some twenty minutes later. The salmon fishers who spread their nets here in Summer cannot tolerate the very expensive damage caused by seals, and will shoot on sight.
~ By contrast, the common seals in Oban bay who enjoy the unwanted fish thrown to them by fishermen, are not persecuted, and show no fear of humans.

FEBRUARY

~ DOWN
BY THE
ROCK POOLS ~

February

Ceud mhios an earraich ~ The first month of Spring

~ February is sometimes called ~
the gate of the year.

'Tis February's changeful mood
When eve to morn is seldom true
And day which broke gusty and rude
Oft shuts in skies of softest hue
In mild repose one sun goes down
The next comes up with murky frown
But scarce hath tolled the hour of day
When glittering roll those frowns away
~ Caroline Webb ~

~ February brings the rain
Thaws the frozen lake again ~
~ Sarah Coleridge ~

Feb 1.

Bright cold weather has followed a thaw of lying snow, and crossing the ferry in good time we enjoyed a brisk morning walk along
the Seil coastline as far as Cuan Sound. The track we followed began
at a kissing-gate and ran between the shore and the cliffs. The gate
looked as if it bore the marks of many years of weathering by fierce
elements. An almost vertical fall of water off the hills fed a small
burn which meandered among the rushes. To avoid the marshy ground
we followed the rocky high-water line and came across a dead seal
half hidden in a bank of seaweed. The poor creature was a half-
grown grey, and the mottled markings of its underside were clear to
see.   Near the strait that separates the isle of Luing
from Seil we watched a cormorant which appeared
to be swimming backwards! The current at
this point swirls through at ten knots and
the bird could make no headway against it.

BARN OWL

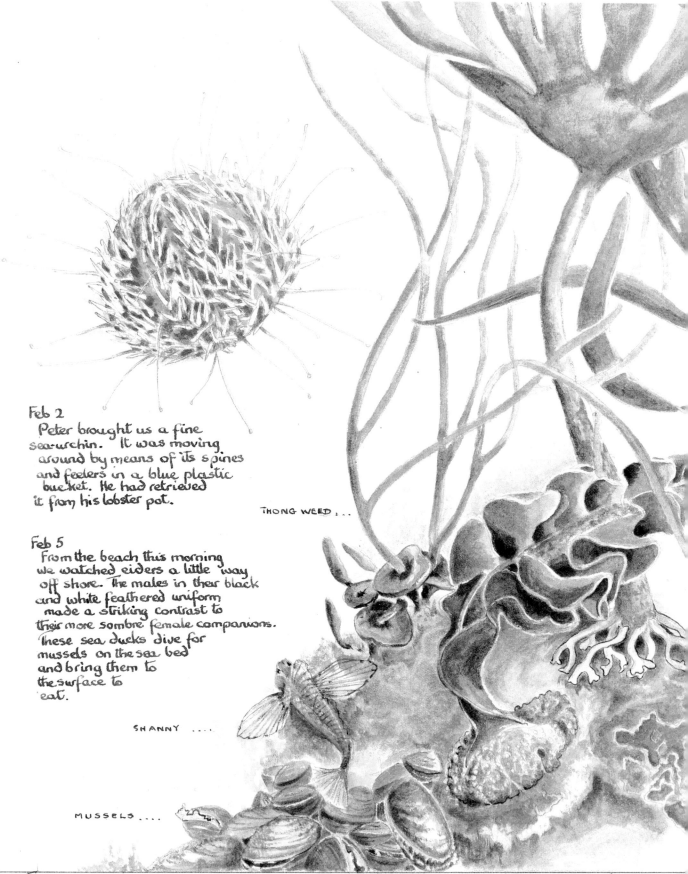

Feb 2
  Peter brought us a fine
sea-urchin. It was moving
 around by means of its spines
and feelers in a blue plastic
  bucket. He had retrieved
it from his lobster pot.

Feb 5
  From the beach this morning
we watched eiders a little way
off shore. The males in their black
and white feathered uniform
  made a striking contrast to
their more sombre female companions.
These sea ducks dive for
mussels on the sea bed
and bring them to
the surface to
eat.

THONG WEED ...

SHANNY ....

MUSSELS ....

The larger lower shore seaweeds that grow here are mostly oarweeds and furbelows. The stems of

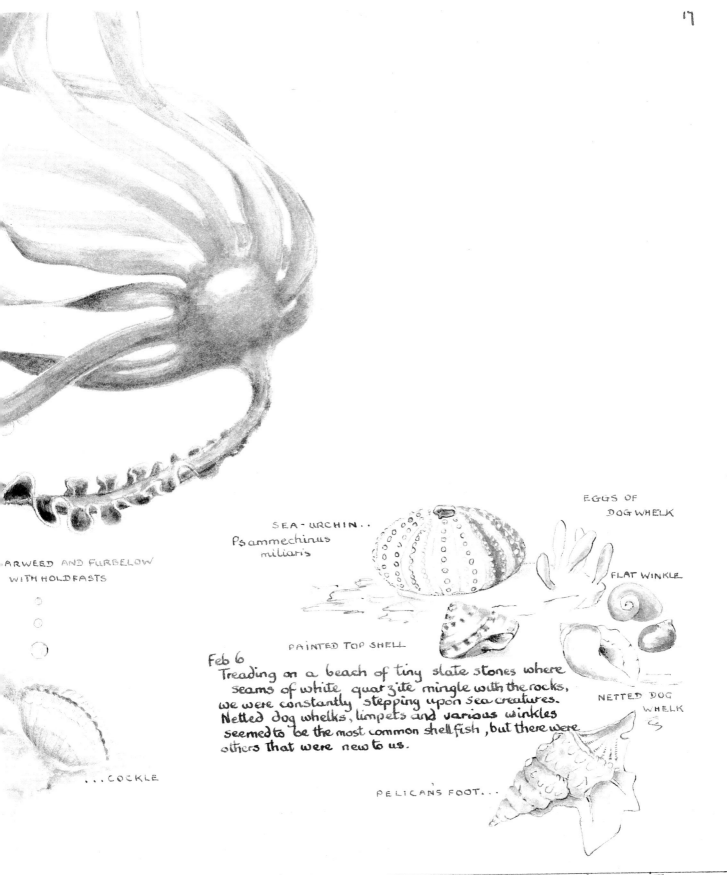

OARWEED AND FURBELOW
WITH HOLDFASTS

...COCKLE

SEA-URCHIN..
Psammechinus
miliaris

PAINTED TOP SHELL

EGGS OF
DOG WHELK

FLAT WINKLE

NETTED DOG
WHELK

PELICANS FOOT...

Feb 6
Treading on a beach of tiny slate stones where
seams of white quartzite mingle with the rocks,
we were constantly stepping upon sea creatures.
Netted dog whelks, limpets and various winkles
seemed to be the most common shellfish, but there were
others that were new to us.

oarweed measure as much as 15cm in circumference and are strewn on the beach with
their holdfasts wrenched from their rocky anchorages in the recent gales.

Feb. 14 St. Valentine's Day

In a favourite old book of ours, yellowed and much thumbed since the days of its first owner (who inserted the date 1893), is a quotation which reads thus ~ "The fourteenth of February is a day sacred to St. Valentine. It was a very odd notion alluded to by Shakespeare that on this day birds began to couple: hence perhaps arose the custom of sending on this day letters of love and affection".

Delving into this book in particular we have found scraps of prose and poetry which though gleaned from another era seem to marry well with the diary. We feel they bring a certain freshness and sometimes stillness from a less turbulent age.

. . . . .

My true love hath my heart and I have his
By just exchange one for another given
I hold his dear and mine he cannot miss
There never was a better bargain driven
My true love hath my heart and I have his

~ The Bargain ~
~ Sir Philip Sidney ~

There was a wren on the shore this morning, her breast reflecting gold in the sunlight, pecking for insects among the crevices as she hopped from rock to rock in her mouselike way. A last-year's-nest is tucked among tussocks on the vertical face of the ridge surrounded by primrose leaves. We wonder if she raised a family there last Spring.

. . . .

Feb. 15

Facing Mull is one of our favourite spots where one can imagine that a giant once enjoyed himself flinging huge boulders about. Among them are fascinating rock pools the like of which we have never seen anywhere else. One in particular contains among the limpets and small seaweeds a lovely specimen of the dahlia anemone. It is more than ten centimetres across when fully extended and its ten-

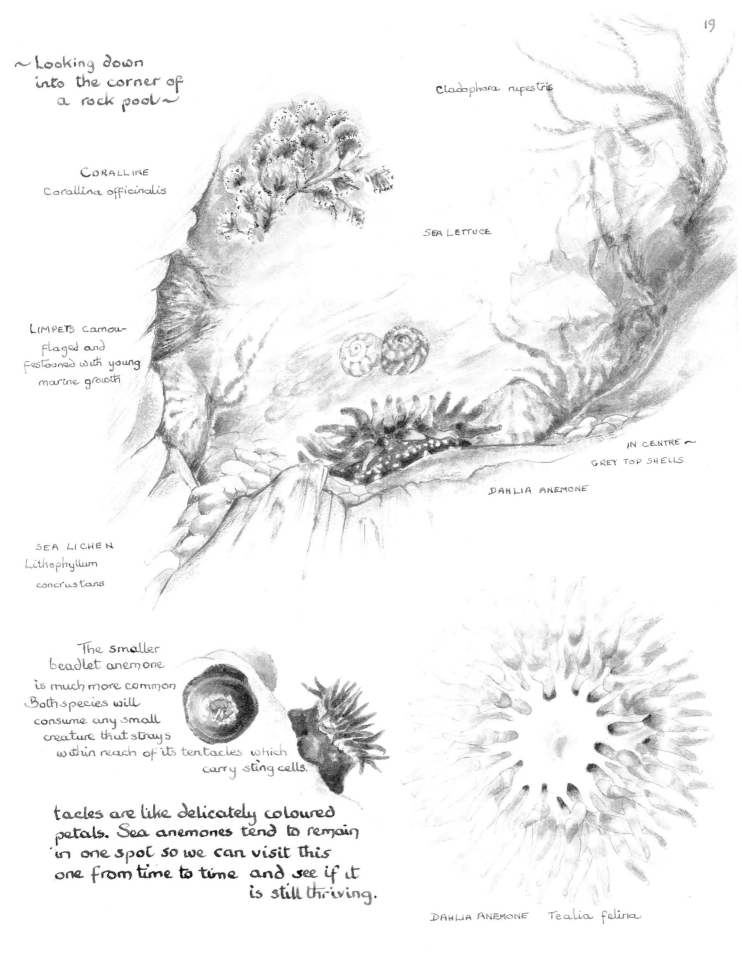

~ Looking down
into the corner of
a rock pool ~

Cladophora rupestris

CORALLINE
Corallina officinalis

SEA LETTUCE

LIMPETS camou-
flaged and
festooned with young
marine growth

IN CENTRE ~
GREY TOP SHELLS

DAHLIA ANEMONE

SEA LICHEN
Lithophyllum
concrustans

The smaller
beadlet anemone
is much more common
Both species will
consume any small
creature that strays
within reach of its tentacles which
carry sting cells.

tacles are like delicately coloured
petals. Sea anemones tend to remain
in one spot so we can visit this
one from time to time and see if it
is still thriving.

DAHLIA ANEMONE   Tealia felina

Feb 16

A bright morning of hard frost
Saw the single plover again this
morning strutting about on a
little cliff edge facing the sea. We
were able to watch him for a while,
as although he knew of our presence he seemed in no hurry to
leave. It would please us if this
pretty little bird should breed
here during the Spring.

A cultivated strain of snowdrop
is growing under the little willow
tree at the corner of the lawn.

Feb 19

While Vicky was sketching at Kilbrandon I made my way to the kirk
and tried the single-manual organ. The church yard seemed to have an
attraction for moles which had been busy sending up their heaps of fine earth
beneath the sitka spruce. Many snowdrops are blooming in the cottage
gardens and at Kilbrandon House.

Feb 20

We were amused this morning. ~ A few hens are kept by a neighbour,
and these run freely about the upper shore. A herring-gull, fancying a scrap
it had spotted among the grass strayed into the territory where the hens
were feeding. The indignant and aggressive reaction of one of the hens
came as a surprise to the gull which retreated inspite of its size. The
same gull wakes Grace, our neighbour, impatiently reminding her to attend to its breakfast.

Feb 23

With the barometer at its highest recording this year, 1051 millibars,
the day was springlike and sunny from dawn till dusk. The weather
raised the gnats which were dancing. There is a special freshness
about the mosses and ferns whose greenness is a pleasant change from
the tired grass. On some parts of the island the grass has been burnt
off to stimulate new growth. Brought home some of the lovely moss
which carpets the boulders at the foot of the ridge.

Feb 24

The food put out on the window sill does not now disappear so quickly. The chaffinches and starlings have ceased their visits and the dunnock and robin come less often than they did when the deep snow was about. The song-thrush and blackbird still enjoy a meal at the sill and both have a delightful habit of sitting motionless on the back of the garden seat as if in deep meditation. The thrush has chosen the cottage roof as a singing post and is now in full song. This begins at twilight each morning and is heard again before sunset. The cock blackbird has begun a tentative sub-song at present so quiet it is drowned by both the dunnock and the robin.

Feb 26.

Against a patchwork of bramble and dead fern in the shadow of the ridge, I startled a bird which rose almost vertically and was quickly out of sight.

From the colour of its plumage and the chequered pattern I could see it was a woodcock.

The bird is said to carry its young between its thighs when in flight.

Feb 27

At Dun Mor one can clearly see four mountain goats on the preci- pices of the cliff face. With them is a young kid now a month old, and born on a rough night in January. The choice of such an insecure spot seemed extraordinary for the birth, but these sure-footed crea- tures seem to know what is best for them.

We are fond of this poem by Browning.

Grow old along with me,
The best is yet to be,
The last of life, for which the first was made;
Our times are in his hand
Who saith 'A whole I planned,
Youth shows but half; trust God: see all, nor be afraid!'

RABBI BEN EZRA ~ Browning

Feb 28

As if to remind us that 'our times are in his hand,' and that the progression of the sea- sons can not be hur- ried, the barometer fell sharply to-day. Its warning was pre- cipitated when sleet and snow suddenly hit us. We were in Kilbrandon church sev- eral hours later. The sun was streaming through the lovely stained-glass windows. On the steep steps outside, we found a barn owl feather.

MARCH

~ CLACHAN
BRIDGE,
SEIL ~

March

Am mart ~ Seed time

~ February makes a bridge
but March breaks it ~
G Herbert

~ The lengthening month that wakes
the adder and blooms the whin ~
Norse folk-lore

This is the month we dock the night
Of a whole hour of candlelight
When song of linnet and thrush is heard
And love stirs in the heart of a bird.
K. T. Hinkson.

LINNET
and
GREENFINCH

Mar 1

Among the creatures that are not
found on the island is the adder.
In fact there are no snakes, rabbits
hedgehogs or mice.
The highest point on Easdale is only
thirty-eight metres, but this is high
enough to give a bird's eye view in all
directions for many kilometres.

. . . . . .

The ridge which runs from north-west to
south-east is easily accessible by a little
path which skirts the rockface at the
north end. ~ It will soon be embroidered
with a mass of Spring flowers.
Nesting birds, including greenfinches,
sometimes called green linnets will inhabit the scrub and the ivy-mantled
ledges that face the south. Occasionally
the ferryman's white domestic goat is tethered where it can clamber on to the
skyline. Its owner needs to be agile to climb after it and bring it into shelter
for the night. The two kids that used to accompany her were sold in January, but
we hear she is expecting again.

From Field Notebook

Rosie of the Ridge's southern aspect

Black Guillemot

We watched him through the
binoculars, from the derelict
pier at Ellenabeich

Mar 4

For several weeks we have seen a single black guillemot near the old wooden pier. Although it is early in the year this bird is in its summer plumage, a coal black relieved only by a white wing patch. When it alights its bright red legs are easily noticed.

Gulls appear to be splitting into pairs. A couple of common gulls were sitting close together on a rock near the water's edge. Herring gulls are behaving similarly and seem to have become more noisy.

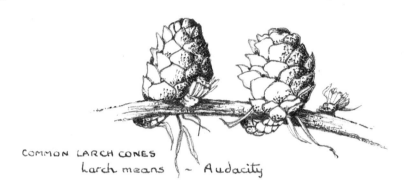

COMMON LARCH CONES
Larch means ⟨ ~ Audacity

Mar 9

Went to Kilninver and Seil in search of signs of Spring. Hazels in the hedges bore male and female catkins and the sallow was bedecked with furry silver knobs. Brought home some sprays of larch bearing last years cones; they look well in the apostle jug. At the water's edge at Loch Feochan an area of flat rock was littered with smashed shells of cockles, mussels, whelks and both Portuguese and common oysters ~ obviously the prey of waders.

Mar 10

Back home, minute buds are beginning to form on the blackthorn. Among these bushes are thrushes, blackbirds, dunnocks, greenfinches and wrens. They are seen flying to and fro surveying their territories. A cock reed bunting is often seen and heard now at his favourite perch in the rowan tree in our garden. ~ Buds are swelling on this tree.

The year's at the spring,
And day's at the morn;
Morning's at seven;
The hill-side's dew-pearled
The lark's on the wing
The snail's on the thorn
God's in his heaven
All's right with the world.

R. Browning

Meanings
Ivy   ~ Fidelity in friendship
Hazel ~ Reconciliation
Willow ~ Freedom

IVY,
CURLEW,
HAZEL, and WILLOW
              (Sallow)
   CATKINS.

~ Blossom ~

The nightingale
has a lyre
of gold
The lark's
is a clarion call

And the blackbird
plays but a box wood
flute
But I love him best
of all
~ William Ernest
Henley ~

HEN BLACKBIRD

Mar 12

Although the frost has gone, the friendly hen black
bird is still with us. We have become very fond of
her and have named her Blossom. Sometimes on a
fine day she will appear, her brown plumage tight
to her body and with a soft bloom upon it; on colder
days she puffs out her feathers to trap a greater a-
mount of air to keep her warm. The two cock birds
which have visited us are strikingly handsome with
sleek black feathering contrasting boldly with their
orange beaks and circles around each eye.
In song, the blackbird is loud, melodious and flute-
like, each phrase different from the preceding, its
range considerably more than an octave.
It is only the cock-bird that sings.

~Friendship is the inexpressible comfort of feeling safe with a person, having neither to weigh thoughts nor measure words. ~

~George Eliot~

~ Ivy means friendship ~

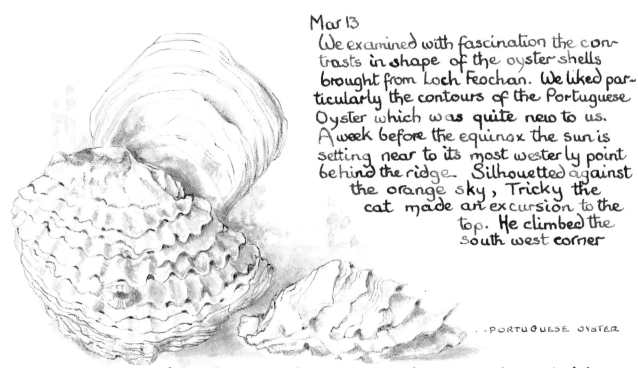

Mar 13

We examined with fascination the con-
trasts in shape of the oyster shells
brought from Loch Feochan. We liked particularly the contours of the Portuguese
Oyster which was quite new to us.
A week before the equinox the sun is
setting near to its most westerly point
behind the ridge. Silhouetted against
the orange sky, Tricky the
cat made an excursion to the
top. He climbed the
south west corner

· PORTUGUESE OYSTER

and soon appeared on the crest, his tail carried high as he picked his
way confidently. The sight proved too much for a couple of hooded
crows which showed their displeasure by dive-bombing the cat and
cawing indignantly. Tricky has a strong personality and is fond of
peppermints!

Mar 15

A delivery of shrubs arrived to-day, and the continuous rain was un-
comfortable to work in. We now have a Japanese Maple on the lawn
at the back of the cottage, and against the slate slab wall at the south
side, a Magnolia soulangeana and Buddleia davidii. On the opposite side
are Pyracantha and Forsythia. Between them, these should please the bees
and butterflies.
On our little plot at the library cottage we have prepared another kitchen
garden. We are already wondering how we shall keep the wandering
donkey out of the plot which is not fenced.

Mar 17

To night we put the clocks forward one hour. Nearing the Vernal Equinox
the longer daylight hours are stimulating activity among the birds. To-
day a pied wagtail rose from the path, while overhead a sky lark at-
tempted a trial stanza from his aria. The song thrushes are in full
voice, and the dunnocks are flapping their wings in courtship display.
The blackbirds seem reluctant to sing.

Mar 21   On the first day of Spring, Vicky has prepared her flower-
press, and the search is on for early blooms. We intend to make a collection

To a sleeping violet ~

Oh Spring, I know thee!
    Seek for sweet surprise
In the young children's eyes.
But I have learnt the years,
    And I know the yet
    Leaf - folded violet
                ~ Alice Meynell ~

. . . . .

Such a starved bank of moss
    Till that May-morn
Blue ran the flash across:
    Violets were born
                ~ Browning ~

. . . . .

The sweetness of the violet's
    Deep blue eyes
Kissed by the breath
                of heaven
    Seems coloured by
                its skys
                ~ Byron ~

. . . . .

of as many flowers as possible from the island. Sure enough this morning I found one single celandine, but when I fetched Vicky to share my delight an hour or two later, the flower had gone. The goat had been tethered close enough for her to swallow it with a mouthful of grass. Further on in the shade, a solitary primrose was brightening the scene, and on a rocky patch in the hollow we discovered a clump of early forget-me-nots with some bittercress.

March. 31.

On the spray-drenched rocks at the west point of the island, a pair of greater black-backed gulls sat gazing out to sea. They seemed to be interested in the lobster fisherman's yellow boat as it ploughed its way through a squall. The distant peaks of Mull suddenly became obscured as a snow shower was drawn like a white net curtain across the scene. The yellow boat appeared ghostly as it was swallowed up in the storm cloud.

The buds on the larch sprays have sprouted very quickly in the warmth of the living-room. They remind one of mini shaving brushes.

APRIL

THE LOBSTER FISHERMAN

April
~ An Giblean. ~

~ The smiling Spring comes in rejoicing ~
And surely Winter grimly flies ~

~ April, April,
Laugh thy girlish laughter;
Then, the moment after,
Weep thy girlish tears! ~
Sir William Watson

DUNNOCKS

~ When proud-pied April, dress'd
in all his trim,
Hath put a spirit of youth in everything ~
Shakespeare

Apr 1

Lesser celandines are starring the west facing banks of the ridge. Under the canopy of bramble they are sheltered from the wind. On this side it is noticeably warmer than the bleak exposed eastern aspect.

Apr 3

The strong north wind was slicing off the tops of the waves to-day before they could break, and dispersing them in a cloud of spume. Dunnocks have for some days become more excited in their wing flapping courtship displays. There are several pairs on the island and we shall look out for their nests.

At first light two common gulls were pacing the turf and enjoying a breakfast of worms. We have never seen the black headed gulls-which are our favourites-on Easdale Isle, though we often share a packed lunch with some by the side of Loch Feochan, when driving home from Oban. The celandines flowering at first on very short stalks are growing fast. I spotted a red-throated diver not far from the ferry. It will be late in the month, or not even until May before the black headed gulls start breeding.

LESSER CELANDINES
Ranunculus Ficaria

Pansies, lilies, king-cups, daisies
Let them live upon their praises
Long as there's a sun that sets
Primroses will have their glory
Long as there are violets
They will have a place in story
There's a flower that shall be mine
~ 'Tis the little celandine ~
~William Wordsworth~

~To the same
flower ~

Pleasures newly found are sweet
When they lie about our feet
February last my heart
First at sight of thee was glad
All unheard of as thou art
Thou must needs I think have had
Celandine! and long ago
Praise of which I nothing know
~William Wordsworth~

Celandine means ~
Joys to come.

When fishes flew
     and forests walked
And figs grew upon thorn
Some moment ~
when the moon was blood
Then surely I was born.

The tatter'd outlaw
                    of the earth
Of ancient crooked will
Starve, scourge, deride me
                    I am dumb
I keep my secret still.

With monstrous head
     and sickening cry
And ears,
          like errant wings
The devil's walking parody
Of all four-footed things.

Fools! For I also
                    had my hour
One far fierce hour; and sweet
There was a shout
                    above my ears
And palms before my feet.
                    G. K. Chesterton.

Apr 6   Palm Sunday
Found Vicky sitting on the bank
painting celandines. The sun was intensely
bright and the picked blooms were
casting deep shadows on the paper.
Annabella the donkey has free
range over the island: she is inquisitive
and will quietly overtake one who is walking,
giving a friendly nip to the coat or shirt. She stands
watching her reflection in the cottage windows.
To-day she approached the headland, and stood gazing
out towards Mull as if wondering if others of her
kind resided there. Mr B, our neighbour whose garden
comprises only two daffodils, lost one of them to-day.

MEADOW PIPIT'S NEST....

~ We hope to find one later in the season ~

Apr 7
The island is attractive to both meadow, and rock pipits. Just now the former are engaging in nuptial display. The cock uttering his plaintive call, seek, seek, seek rises lark-like to about twenty metres and parachutes to the ground with an ever-quickening chirp, sounding rather like a clutch of day-old chicks. The rock pipits stay on the shoreline and are nervous of close approach.

Apr 8
The island toads which hide away during the winter (I unearthed one deep inside a slate slab wall I was demolishing at Christmas) have made their annual spring migration to their favourite ponds. There are only two little pools that are suitable. Here they are coupling feverishly and the strands of spawn look like pairs of black leather bootlaces tangled among the water weeds.

Apr 9
The toads continue their aquatic breeding, and there are occasional faint croaking sounds something like a distant crow calling. We have not yet seen any frogs here, but on Seil we came across a tiny pool fed by a spring on the hillside. The water was crystal clear and contained frog spawn and some free-swimming tadpoles. I brought about twenty five back home and shall attempt to rear them in a small aquarium. Returning to the subject of toads, Vicky did not take too well to them at first, but is rapidly developing an affection for these gentle creatures. Perhaps realizing that they are not slimy to handle... has brought about this change of heart.

Apri 10
The frog tadpoles have settled down well. They have natural pond water and weed, and their diet is supplemented with chopped worms and tiny pieces of raw meat. They take this readily, and also pick up and swallow cheese crumbs. If they survive I hope to establish a colony on the island.

TOADS in amplexus (Mating)

BLACK-HEADED GULLS

Apr 13   Good Friday

For several weeks now we have watched with interest clumps of small bright green leaves growing close to the ground near the high water line. Some are round and some heart-shaped, all thick and shiny as if varnished.
The southern end of the island is not one of our favourite spots, but wandering there we were surprised to find more of these plants at an advanced level of development. Their tiny sprays of white flowers were seen a-mong leaves, many of which had changed to lovely shades of yellow orange and plum.

. . . .

Magnified
flowers of.
SCURVY GRASS

Under magnification the sweet smelling flower *Cochlearia officinalis* (one of the CRUCIFER family) has four white petals arranged in the shape of a cross. Popularly known as 'Scurvy grass' it is so called because of its use at one time as a herbal remedy for the sickness of the same name to which sailors were prone, Salted provisions with a lack of fresh fruit and vegetables on long journeys at sea being the cause.

Apr 15   Easter Day

Our cottage door has no slit for letters, so the postman puts the mail in a wooden box on the floor of the porch. This morning I glanced into the box on leaving and found a toad sitting there. I noted her attractive colouring of cream, tan and brown. On releasing her in the garden she immediately began burying herself. A blackbird's nest among the brambles just beyond our garden wall contains one egg. A peacock butterfly out of hibernation inspected the rockfoil to-day.

V. Waite

A few leaves of Coltsfoot
are thrusting their yellow
tipped spears through the
slate rubble.
Normally the plant enjoys
heavy land, and is called
Clay leaf in some places.
Also known as Cough wort
it was used medically
for chest complaints.
The name Coltsfoot derives
from the shape of the leaf
which appears later.
Close to the ground, the
leaf looks rather like
the imprint of a horse's
hoof.

Apr 16 Easter Monday.   Fine weather.
A small tortoiseshell butterfly came into the garden to-day.
The fields on Seil are dotted with tiny lambs, which show up clearly
against the pale green of the new grass. Most of them are like
white dots, but among them are a few black ones. Rarely does one see
a fully grown black sheep; their fleeces are not marketable, and so
such lambs must go to the butcher.

Apr 18
The whistling cry of the oyster catcher is frequently heard. These
pretty birds are often seen on the rocks. Their red feet
must be perpetually wet. At rest they
turn their heads backwards, and
at a distance appear to have
no bill. They are curiously
named; some will rarely
see an oyster. Mussels are
most commonly devoured.

Apr 21.
On the same day that the first blackthorn was in bloom, a small
white butterfly was on the wing. Its visits will be less welcome
when there are cabbages growing in the garden.
A flock of curlew flew over, their musical cries echoing round the island
We found the first violets in the Donkey hollow, thus named because
we discovered Annabella for the first time there.

Apr 23
On an island that is almost treeless one would hardly expect to see and
hear members of the leaf warbler family; but to-day I rested on the lawn
eyes closed, concentrating on the quietness, and deliberately noting
every sound. Quite suddenly I heard the delicate rippling cadence of
the willow warbler's song. No sooner had it registered, than the little bird
entered the garden and sang immediately over my head in the Rowan.
My eyes were now wide open, and I watched the warbler fly off again.

Apr 26
Not far from the cottage door I picked up a starling's egg, slightly chip-
ped, but otherwise whole. It lay on the ground just below a crevice under
the slates in the next cottage where a starling is nesting. There was no
evidence that a predator had been at work. Perhaps the
hen bird had ejected the egg involuntarily
while feeding before she had time
to return to the nest.

~ The birds will be breeding in June ~
Egg approx. 4/5^{ths} life size

..Shell of a SPINY SPIDER CRAB

Apr 29

After a lapse of many weeks we decided to see how "Fred" (our name for the Dahlia anemone) was faring. He was still securely fastened to the wall of the rock pool where we first discovered him. Since our earlier visit we noticed that the limpets whose shells had been adorned with little plumes of seaweed, and looking like decorated hats, were now sporting long streamers, such was the growth of the seaweed. Found many crabs. Following our individual pursuits around the shore, we often lose one another. I can usually be located by my orange woolly cap, and Vicky manages a particularly penetrating call.

..SHORE CRAB

Apr 30

On turning over a few rocks on the west side at low tide, we uncovered in the space of a few minutes a common starfish, two sand eels, a blenny or two and two brittle stars. Both the starfish and brittle star are five pointed, and have a radial symmetry, but they are quite unlike in their means of locomotion. The starfish moves slowly in any direction by the action of its hundreds of tube feet.
The brittle star has a muscular control of its arms which it lashes about, enabling it to move much faster. Their brittleness causes them to break easily. The larger specimen we found had suffered some damage.

..BRITTLE STARS

..... PORCELAIN CRAB

MAY

~YOUNG CATTLE
NEAR ARDENCAPLE~

May

An céitean ~ Beginning of Summer.

~ Such a starved bank of Moss
Till that May ~ Morn
Blue ran the flash across
Violets were born ~
Robert Browning

~ Ne'er cast a clout
Till May be out ~
Fuller

~ A CUP OF THRUSHES ~

~ May brings flocks of woolly lambs
Sporting round their fleecy dams ~
Sarah Coleridge

May 1

A day of snow and hail showers did not prevent our walking on the wet grass and in the cold wind. Violets and primroses carpeted the sunny slope of the ridge, their petals shimmering with shower droplets. The whimper of a very young lamb came to us across the still air from Dunmor farm on Seil.

May 2

The cold weather persists. Blossom has five eggs in her mossy nest, and they are due to hatch any day now. A buzzard overhead was mobbed by angry gulls whose aggressive instincts were sharpened by the breeding season. There is a thrush's nest in the rhododendron near the church path. It is well screened, and contains four tiny babies. The parent birds were a little disturbed as Vicky went to work with her pencil, but they soon accepted her. ~ Sunday must be a bothersome day for them!

Lakeland Squirrel
enjoying the
tender young leaves
of beech

~ Benefits
of
Spring ~

There's perfume up on every wind,
Music in every tree,
Dews for the moisture ~ loving flowers,
Sweets for the sucking bee. ~ N. P. Willis

LESSER CELANDINE.
SCURVY GRASS
YOUNG BIRCH SHOOT
OSIER CATKINS
DOG VIOLETS
BUFF TAILED HUMBLE BEE

It wins my admiration ~
To view the structure of this little work ~
A bird's nest.
     ~ Mark it well, within, without,
No tool had she that wrought ~
     no knife to cut,
No nail to fix, no bodkin to insert,
No glue to join, her little beak was all
And yet how neatly finished ~
     what nice hand,
And every implement and means of art
And twenty years apprenticeship to boot
Could make me such another?

     Fondly then
     We boast of excellence,
     whose noblest skill
     Instinctive genius foils.

~ A Bird's Nest ~
Hurdis.

Blackbird's
Egg

(Life size)

~ BLACKBIRD'S

NEST ~

~ May 3 ~

Having tended about twenty-
five frog tadpoles since collecting them
as eggs late March, it is good to note they
are all thriving and growing. Their progress has
been fascinating to watch in the aquarium which
is their home. From little black wrigglers of less than
a centimetre in length they have developed into vigorous fatties of nearly three centimetres (two thirds of which
is the length of their tails). Each of them now has the twin
buds which will grow into hind legs. They have been fed
on pond-weed and chopped worm.~ Already their bodies
are frog-shaped.

~ May 4 ~

While the frogs have depended on daily attention in
our domestic environment, an island pond has contained
a much larger quantity of toad tadpoles. These emerged
from eggs laid in early April when dozens of toads
congregated. During their nuptial embrace strings
of spawn were twined among the weed. They have
now become tadpoles very similar to those of the
frog, but at a much earlier stage of development. The same pond holds a few smooth
newts, but if they breed they will be
outnumbered by a hundred to one.
~ No doubt the heron knows
where they are......

Speak, whimp'ring younglings, and make known
The reason why
Ye droop and weep;
Is it for want of sleep,
Or childish lullaby?
Or that ye have not seen as yet
The violet?
Or brought a kiss
From that Sweet~heart, to this?
~ No, no, this sorrow shown
By your tears shed,
Would have this lecture read,
That things of greatest, so of meanest worth,
Conceived with grief are, and with tears
brought forth.

~To Primroses filled with Morning Dew~
Robert Herrick

Enlarged
flower head...

WILLOW WARBLERS
and
EUROPEAN LARCH
Larix decidua

WOOD SORRELL
Oxalis acetosella

WOOD MOUSE....
and WILD HYACINTH—
Scilla nutans

Wee sleekit, cowrin tim'rous beastie
O, what a panic's in thy breastie!
Thou need na start awa sae hasty
      Wi' bickering brattle!
I wad be laith to rin an' chase thee,
Wi' murd'ring pattle.
~ Robert Burns ~

~Seil~ In the forestry, and at Ardencaple

... CAMUS LAICH

May 5

A day on Seil. On our way to a forestry plantation to see and hear the willow warblers, we noticed that the pretty pollen-speckled male catkins of the osier were now accompanied by new leaves. We went on to picnic near Ardencaple, and sitting very still, we watched a woodmouse attending to his toilet. Suddenly he became aware of us and bounded away in little jumps among the wood sorrel. The larch trees were flowering delicately.

> I have looked o'er the hills of the Stormy North
> And the larch has hung all his tassels forth.
> ~ F.D. Hemans ~

At Ardencaple, moss clad rocks clasped by birch roots give way to water-meadows which meet the sea, and Ben More on Mull in the distance was first clear, then blotted out, as snow flurries passed over the peak. When a snow shower came across in our direction we sheltered in a derelict farmhouse much used by cattle and sheep in very severe weather. A swallow's nest in the rafters reminded us that the birds would soon be returning. Looking up as we left the ruin, we noticed the sun now revealed, was ringed with a bright light, and across at the water margin two red deer hinds were grazing. On our way back crossing marshy ground carpeted with sphagnum moss we came across a roe deer lying dead in a ditch. There seemed to be no mark of injury on the animal, which must have been alive not many hours before. In death it was still beautiful.

May 6
Blossom and her husband are busy gathering worms for their first brood. Working from dawn to dusk for their hungry family there is scarcely time for preening themselves, and both are looking rather tatty.

May 7
A visit to the mainland. On a trip to Arrochar and Tarbet, we saw a buzzard sitting on a roadside post. Quite heedless of the traffic it appeared preoccupied with private thoughts. In a nearby field two mares with day old foals were quietly dozing. Our return journey took us back by Loch Awe where daffodils decorated the roadside.

May 12
At last a couple of swallows have arrived. A moist warm day brought them skimming at low level above the blooms of sea pinks on the bare rocks. Ramsons and lady's fingers are blooming, and the ling heather is showing new growth. There is an old saying ~

~ Eat leekes in Lide (March)
And ramsons in May
And all the year after
Physicians may play ~

Small white butterflies are on the wing. They have already laid eggs on brassicas and nasturtiums and the first caterpillars have begun feeding. The frog tadpoles have been given their freedom. Their progress will be watched in the small pond halfway up the ridge. Found a Meadow pipit's nest with four eggs deep in the grass.

May 15
I keep the grass in front of the cottage very short, and just now it is covered with common daisies. I watched a pair of twite tearing away the pink-tipped petals, and feasting on the gold centres. A little later two pied wagtails seeming to find the close cropped grass to their liking, were busy scuttling about with tails a-wagging.

Pluck not the wayside flower,
It is the travellers dower,
Hundreds in passing by
May espy...
Then spare the wayside flower.
Allingham

May 17 On a journey to Moffat in Dumfriesshire we drove through that kind of Scottish weather which keeps the burns flowing, and the lochs filled. The wooded grounds of Inverary Castle rang with the singing of

WOOD ANEMONE ~ Anemone nemorosa

As in the sunshine of the morn
  A butterfly, but newly born
    Sat proudly perking!

And what's a butterfly?   At best
  He's but a caterpillar drest:
                        ~ John Gay ~

        Lovely light airy thing
        Thou butterfly.
                    ~ Herder ~

~ BUTTERFLIES ~

SMALL WHITES
   (Courting)
      and
GREEN VEINED WHITE

      on

   THRIFT

Blackthorn
Prunus spinoza
~ Difficulty ~

Thrift
Armeria maritima
~ Sympathy ~

Ramsons
Allium ursinum

~ Spring, the joyous Spring, has come;
The flowers awake, and insects hum;

The prickly sloe is decked with white,
The blackbird sings from morn till night.

The gale has died; a gentler breeze
Is rustling in the rowan trees.

The toad has left his Winter den
And song bursts from the bustling wren.

~ Branch of Rowan ~ Small Tortoiseshell butterfly

The saplings shoot from every stool
And celandines surround the pool,

The early waking butterfly
Forsakes the dark and seeks the sky.

All hail great spirit ~ nature's king
Whose subjects breathe the breath of Spring~
~G. Waite~
(When a schoolboy)

Blackthorn blossom (Sloe) ~ Celandines.

chaffinches and willow warblers and we heard the first cuckoo of the year. The rain imparted a delightful freshness to the newly opened beech buds from which drops fell upon wood sorrel. Decaying stumps were clothed in luxurious green moss, and between them cream and pink sprays of rhododendron were bursting into fullness.

Having parked the caravan at a farm just outside Moffat, we were lulled to sleep by the voices of ewes on the hillside calling in their lambs for the night.

May 18

A brief stop on the roadside by the banks of Ullswater gave us the unexpected sight of a lovely red squirrel, which took little notice of us as he selected tender beech buds from the slenderest twigs. The quickly spreading canopy of vivid green beech leaves formed a screen for the little fellow as he moved swiftly from branch to branch.

We were in the district where burns become becks, and lochs, waters or meres. In the brightness of the morning, we were surprised to hear a tawny owl calling in a roadside wood. Perhaps he was talking in his sleep. If this owl is discovered by small birds in daylight they will recognize him as a predator and mob him. Nearing Windermere we passed a bank of cowslips, less common than they once were. At Grange-over-Sands, we sheltered under some larches where jackdaws splashed about in pools of rainwater brimming from a recent shower.

. . . . .

May 24

There are moments when the
soul takes wings
What it has to remember
it remembers
What it loves — it loves still more
What it longs for — to that it flees. ~

MOUNTAIN EVERLASTING
Antennaria dioica

~IONA ABBEY~

A collared dove in the rowan tree reminded us that tomorrow we go on a much anticipated visit to Iona with our revered friend, M.H. At one time he exchanged letters with the widow of the Celtic writer Fiona McLeod (William Sharp). The above words were written by William Sharp before visiting this Isle to which St Columba brought Christianity to Great Britain. (St Ninian probably preceded him).

May 25

Waiting for the ferry at Fionnphort we had time to gaze out from Mull into a sea of turquoise and sapphire blue. In the evening we went to the Abbey, where one of the white doves alighted on an arm of St. Martin's cross near the entrance.
No wonder the graceful tern is known as the sea-swallow. At the N.E. point on the white Strand of the Monks, groups of them swooped and dived, knitting air with water as they delicately dipped below the water in search of food.

May 26

Our much anticipated search for a certain bay of special stones proved disappointing, but we found several small attractive flowering plants on the way, one, Mountain Everlasting, similar to Edelweiss.
At sunset, returning home from a walk on the white sand, we suddenly found a bay where many stones of lovely colours lay beneath the water. We were loath to leave the place, but returned by way of the machair where cattle were silhouetted against an incredible mixture of rose pink and duck-egg blue which engulfed both sea and sky.

May 27

On the walls of the old Nunnery near the rockery, Ivy-leaved Toadflax garlanded the yellow lichen-covered pink granite walls.

~ Flower in the crannied wall
I pluck you out of the crannies.
I hold you here root and all in my hand
Little flower ~ but if I could understand
What you are ~ root and all,
and all in all
I should know what God and man is ~
TENNYSON

May 28

It was very evident from our vantage point that the turbid mile strip of water between Mull and Iona was a favourite fishing ground for gannets. About twenty of these magnificent birds there were demonstrating their splash dives repeatedly. A change of weather with fresh wind and penetrating rain. This is part of the Scottish scene which lends grandeur and awe to the sight of the mountains. They appeared sombre today as we drove across Mull.
We were happy to find a stowaway in our vehicle, a beautiful green Hairstreak butterfly which made the journey with us, and got out at Craignure.
~ Thumbnail sketch ..

May 31

We have returned from our travels, and so have the swallows.

Having investigated the pleasures of unfamiliar surroundings, we are now back on home ground. An urgent task is the seeking and destruction of eggs of the small white butterfly on the brassica.

Walking on the ridge I found a rabbit skull picked clean by a predator, lying on the turf. It was probably carried there by a buzzard as no rabbits inhabit the island. Descending the slope I felt impelled to lift a flat stone, and there underneath was a toad which inflated itself indignantly. Unmolested it was luckier than another I saw hanging from the beak of a herring gull!

Oystercatchers could be heard and seen performing their courtship display in which the male bowed low before the female, and a Wheatear passed overhead with dried grass in its bill. I wondered what birds other than Gulls, Blackbirds, and Starlings were nesting on the island.

The sun descending in the west
The evening star does shine.
The birds are silent in their nest
And I must seek for mine.
~ William Blake ~

JUNE

THE GREEN OF
INSH ISLAND
FROM
EASDALE'S
ROCKY SHORE

~ROCK PIPIT~

June

An òg mhìos ~ Young Summer month

~ The cuckoo comes in April,
He sings the month of May.
He whistles a tune in the middle of June
And then he flies away ~

Willam Hazlitt

~ June brings tulips, lilies, roses
Fills the children's hands
with posies ~

Sarah Coleridge

BURNET
ROSE

Rosa
pimpinellifolia

The rose of all the World is
not for me ~
I want for my part, the little
white rose of Scotland
That smells sharp and sweet, and
breaks the heart ~

Hugh Macdiarmid

June 1

Mr C, the village postmaster typed out the
poem "The rose of all the world" for us with one finger, having first re-
cited it in the post office. He said it was the first time he had done so without bursting into tears.   Hundreds of tadpoles are active in the toadpool.
The garden is a mass of saxifrage, and cream and pink tulips. The rose
bushes are full of new leaf just now. There is no sign of greenfly, our
worst garden pest being the slug. Rosie the goat, heavy with gestation
is nearing her time. Vicky is busy painting and polishing stones from Iona.

June 3

Brought back a specimen of the lovely early purple orchid that blooms
on the bank of a hollow here. Its stem appears as if stained with magenta.
The pool into which I introduced frog tadpoles seemed to be uninhabited.

Every year the swallows
occupy their mud nests
in a derelict shed on the island...

..PIGNUT
Conopodium majus

~SILVER WEED~
Potentilla anserina

~EARLY PURPLE ORCHID~
) (Orchis mascula)
showing tuber

LOWERING on EASDALE~Dunmor, Seil, in the distance

PIGNUT TUBER cut through

...PIGNUT TUBER

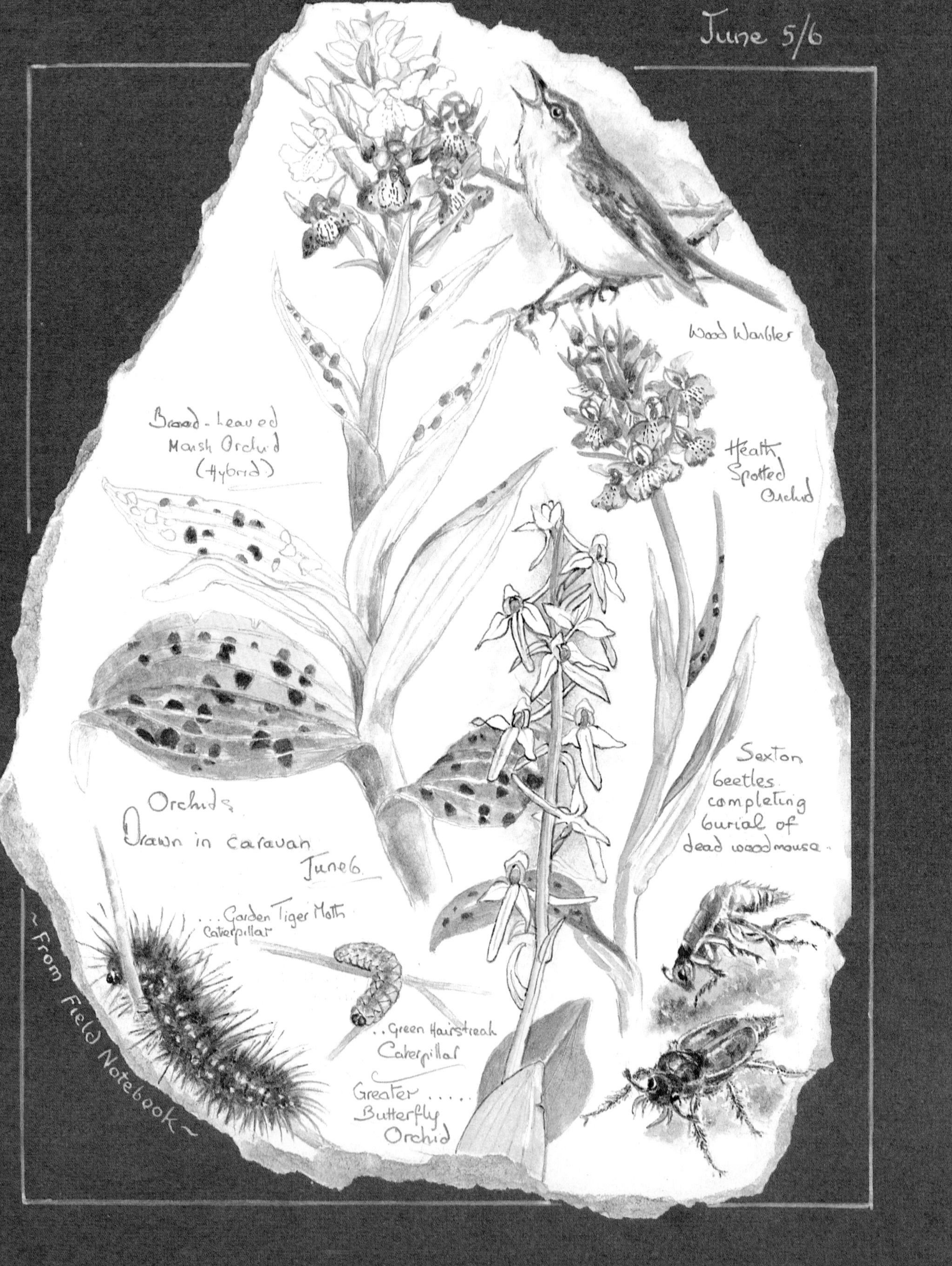
June 5/6
Wood Warbler
Broad-Leaved
Marsh Orchid
(Hybrid)
Heath
Spotted
Orchid
Sexton
beetles
completing
burial of
dead woodmouse.
Orchids
Drawn in caravan
June 6.
... Garden Tiger Moth
Caterpillar
.. Green Hairstreak
Caterpillar
Greater .....
Butterfly
Orchid
~ From Field Notebook ~

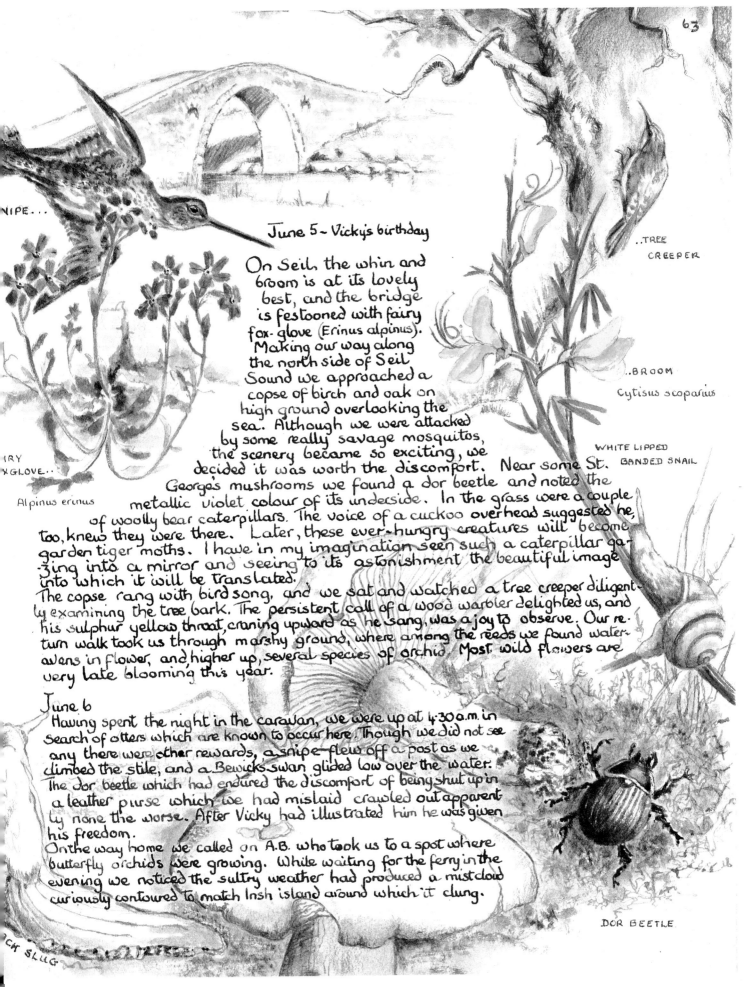

NIPE...

..TREE
CREEPER

..BROOM
Cytisus scoparius

WHITE LIPPED
BANDED SNAIL

IRY
XGLOVE..

Alpinus erinus

June 5 ~ Vicky's birthday

On Seil, the whin and
broom is at its lovely
best, and the bridge
is festooned with fairy
fox-glove (Erinus alpinus).
Making our way along
the north side of Seil
Sound we approached a
copse of birch and oak on
high ground overlooking the
sea. Although we were attacked
by some really savage mosquitos,
the scenery became so exciting, we
decided it was worth the discomfort. Near some St.
George's mushrooms we found a dor beetle and noted the
metallic violet colour of its underside. In the grass were a couple
of woolly bear caterpillars. The voice of a cuckoo overhead suggested he,
too, knew they were there. Later, these ever-hungry creatures will become
garden tiger moths. I have in my imagination seen such a caterpillar ga-
-zing into a mirror and seeing to its astonishment the beautiful image
into which it will be translated.
The copse rang with bird song, and we sat and watched a tree creeper diligent-
ly examining the tree bark. The persistent call of a wood warbler delighted us, and
his sulphur yellow throat, craning upward as he sang, was a joy to observe. Our return walk took us through marshy ground, where among the reeds we found water-
avens in flower, and, higher up, several species of orchid. Most wild flowers are
very late blooming this year.

June 6
Having spent the night in the caravan, we were up at 4.30 a.m. in
search of otters which are known to occur here. Though we did not see
any there were other rewards, a snipe flew off a post as we
climbed the stile, and a Bewick's swan glided low over the water.
The dor beetle which had endured the discomfort of being shut up in
a leather purse which we had mislaid crawled out apparent-
ly none the worse. After Vicky had illustrated him he was given
his freedom.
On the way home we called on A.B. who took us to a spot where
butterfly orchids were growing. While waiting for the ferry in the
evening we noticed the sultry weather had produced a mist cloud
curiously contoured to match Insh island around which it clung.

DOR BEETLE

CK SLUG

Greater Stitchwort
Stellaria Holostea

~June~
Seil

Mayfly

Wild Hyacinth
Endymion non-scriptus
Red Campion
Silene dioica
Cuckoo Flower
Cardamine pratensis

Marsh Marigold.   Wild Strawberry.
Caltha palustris.   Fragaria vesca.

June 7

The sea to-day was like glass, and the islands appeared mystical, suspended in a misty haze.
Sitting on the seat in front of the cottage, it was uncannily quiet. The only sound was of Annabella munching grass. The rowan in the garden is in full flower, and bees and other insects are busy among the white blossom. A small copper butterfly was on the wing to-day. The blackbird's second brood is hatched, and hungry. Tricky, the prowling cat will have to be watched and repelled.

June 8

We found a most attractive flowering rush. Its latin name Luzula sylvatica is said to derive from Luciola ~ a glow-worm, because the dew-drops on the head, sparkle in the sunshine. Watched a pied-wagtail catch a fly on the wing and feed it to a young one.

June 9 ~ St Columba's Feast Day~

Visited Loch Nell. Gathered some fine King-cups (Marsh Marigolds) from the banks of a burn. The blooms of Cuckoo Flower, Greater Stitchwort and Rhododendron were bursting into bloom. The incessant call of the cuckoo was a background to the loud outpouring of the song of a whitethroat. The Primroses and Violets are now joined by the taller spikes of Wild Hyacinth. These Bluebells bloom on the open hillsides, away from any tree cover. In the dry quarry the lovely Burnet rose is growing strongly.

GREAT
WOODRUSH
Luzula sylvatica

WATER VOLE
and HORSE TAILS

Of the three species of vole, the water vole is the largest being about 25 cms. long, (with tail) Bank, and field or short tailed voles are not longer than 13 cms. The former are found on the moors and the latter in plantations.

June 12
A hot day began with a dense mist which completely enveloped the island at sea level. On an afternoon walk, we noticed a bright patch of purple high up on the west of the ridge. Investigation proved it to be thyme in very luxurious flower. On the way up we came upon a common lizard, sunning itself on a stone. It darted away into the grass as we approached.
On our return we turned over a piece of rock to find a nest of red ants underneath. The galleries contained eggs, larvae and pupae, all of which had been carefully stored in batches keeping the queen larvae and pupae separate from the workers.

June 13
Climbed up onto Dunmor rock and went down to a lovely bay where we gathered winkles. Heard the cry of a tawny owl coming from a cliff face, and saw several ravens.

As far as we know, not every wild flower has a meaning, but where we have found one, it is inserted ~

~ Flower Meanings ~
starting from page 1 ~

and in
~ Flower Circle ~ opposite.

WHIN ~ Enduring affection. ANGELICA ~ Inspiration. HAZEL ~ Reconciliation. OSIER ~ Frankness. BEECH ~ Prosperity. COLT'S FOOT ~ Justice. VIOLET ~ Faithfulness. BIRCH ~ Meekness. WOOD SORREL ~ Joy. ROWAN ~ Prudence. WHITE ROSE ~ I am worthy of you. ORCHID ~ A Belle. FOXGLOVE ~ Insincerity. WILD HYACINTH ~ Constancy. MARSH MARIGOLD ~ Desire for riches. STRAWBERRY ~ Foresight. RHODODENDRON ~ Beware. RUSH ~ Docility.

CRAB APPLE BLOSSOM ~ Ill nature. BILBERRY ~ Treachery. MILKWORT ~ Hermitage. CRANE'S BILL ~ Steadfast piety. HEATHER ~ Solitude. GERMANDER SPEEDWELL ~ Facility. FORGET-ME-NOT ~ ~ True Love. BIRD'S FOOT TREFOIL ~ Revenge. CLOVER ~ Think of me.
Also flowering on the island ~
PYRAMID BUGLE. GREY SPEEDWELL. LESSER HAWKWEED ~ Quick-sightedness. ROWAN BLOSSOM ~ Prudence. THYME-LEAVED SPEEDWELL. On Seil, BUTTERFLY ORCHID ~ Gaiety.

SEEDING BUSH VETCH

~Thanksgiving~

Thank you for freedom to wander at will,
To stir into action or sit very still,
The smell of the wild flowers, the hum of the bee,
And silent old islands out there in the sea.

Thank you for giving me skill with a brush
And someone to tell me, ~"There's no need to rush",
The beautiful bird song, and fading of Spring
And just being, able to do one's own thing.

Thank you for tasting, and thank you for sight
And always the peace that one hears in the night,
But most for a partner with giving and sharing
Exchanging a mood for the kindest of caring.

Thanks for soft roses, the clouds and the swallow,
So many thanks ~ yet there's still one to follow;
Thanks for tranquility's farewell to sorrow
And lending me grace, for to-day,
~ and tomorrow~

V. Waite.

CRAB APPLE BLOSSOM · BILBERRY ~ CORNSALAD · POTENTILLA ~ MILKWORT · CRANESBILL · SEA CAMPION
LING HEATHER LEAVES ~ PLANTAIN · GERMANDER SPEEDWELL ~ KIDNEY VETCH ~ COMMON DAISY ·
BUSH VETCH ~ FORGET · ME · NOT · BIRD'S FOOT TREFOIL · CLOVER ~ LING HEATHER LEAVES

~ BLACKBIRD FLEDGLINGS ~

Jun 14

To avoid disturbing her, we have rarely visited the nest where Blossom had her second brood.

Situated in a wall cavity and curtained with young ivy leaves the nest now contains five healthy fledglings, and, to-day, we lifted the curtain so that Vicky could sketch them.

The family appeared as a jumble of beaks and feathers, four of them asleep with beaks in the air. So full was the nest that the fifth was obliged to sit on the rim wide-eyed, and gazing quizzically at us.

While we were there, Blossom arrived with a beakful of worms. At first she waited on a briar, uttering a gentle sub-song, and then ~ satisfied that we meant no harm, came to the nest and fed her young repeatedly.

Jun 19

The salmon-fishers are busy getting their gear ready for the season. My fishing expedition this evening produced only one single coal fish, which I caught on a feather

Two blue butterflies flew into the garden to-day, and settled on some bush vetch the other side of the wall. Thyme, bird's foot trefoil and English stone-crop seem to be everywhere among the slate waste on the island, making natural rockeries. To-day we had rain and sunshine with a succession of rainbows.

Jun. 21 Mid-Summers Day

There is no doubt that various species of birds are nesting on the island; we have seen wheatears, meadow pipits and pied wag-tails carrying food in their beaks, but so far we have not discovered many nests. Sandpipers and greenfinches are showing signs of nesting too, and the cock greenfinches are in splendid colour. In all cases the birds must nest in inaccessible sites in order to escape the dogs that live on the island.

The Viking ship came into view to-day. It was on its way to a rendezvous with a British nuclear submarine.

RHODODENDRON
Rhododendron ponticum

June 25

A beautiful calm evening invited me to take the boat out and make my first circumnavigation of Insh island, two kilometres distant. Although the entire journey lasted only an hour, I felt a curious sensation of being extremely adventurous.

When I reached its far side, I was beginning to feel something like a space man going round the other side of the moon.

On my way out, a common seal popped his head out of the water and eyed me suspiciously. A few minutes later three porpoises broke surface with arched backs. The deep water beyond Insh is the spot in which to look for whales which occasionally follow the current there.

June 26.

To Ellenabeich for messages. (shopping) Irises were coming into bloom at the waters edge.

June 30

We sat for a couple of hours in our favourite secluded spot on the ridge, where there is a commanding view of the North and West sea approaches. Six gannets were fishing very close in-shore; one could hear the plop of their dives quite clearly. A wheatear, quite unaware of our presence was making repeated journeys to its nest in a slate wall below us.

COMMON
SEAL

JULY

DISTANT
VIEW OF
SCARBA
FROM
SEIL

WITH
IMMATURE AND ADULT
COMMON GULLS

July

An seachd-mhios ~ Midmost Summer month

When threatning sky scolds young July
 And thrusts a sudden shower
The slippery sun will surely come
 To laugh away the hour
When silver raindrops spatter grass
 And bid the flowers bend over
In lazy heat we kneel to greet ~
 The sweet warm breath of clover.

V. Waite

WHITE CLOVER among
the slate waste

July 1.
A blustery day with the clouds curtain-
ing the sun at intervals. Looking across
the Firth of Lorn we saw shafts of sunshine
lighting up the white light-house that guards
the shipping lane between Oban and Craignure.
 In the opposite direction, Fladda lighthouse near
Scarba, was glinting, and many gulls took to
 the air as we watched a small dinghy land
on the distant beach. A herring gull allow-
ed our close approach before she flew
from her nest of three eggs.
 The common gulls (less common than others)
nest on inaccessible ledges on a vertical
 face of the ridge, where there is a sheer
 drop to the static water of a flooded
 quarry. The water surface is mirror
 smooth, and reflects the sky in all
 its moods, sometimes sombre and

HERRING GULL

~ Buff ~ and Red-tailed
Humble bees and
Foxgloves

and sometimes even a deep peacock blue. The nests have well grown chicks which have no option but to sit tight on their precarious footholds until their wings are strong enough for flight. Our approach is heralded by a vicious screaming and dive bombing by the parent birds.

————————— • —————————

July 3

We came across a ball of tiny spiders inside their silken canopy attached to a bramble. They looked very much like a seeding flowerhead until disturbed, when they exploded into dispersal. The island grasses are flowering. Among them is a meadow grass which forms a soft blue haze.

On Seil I found some eggs of the two spot ladybird and this evening enjoyed seeing them hatch. I would have liked more time to study them.

On the island we found both the burnet and cinnabar moth, and on Seil, the small heath butterfly ~ (see cover).

July 4

A day on the Isle of Luing. The weather was sunny and warm. The cattle were standing up to their hocks in the sea. No doubt the flick of a wet tail is more discomfitting and deadly to the troublesome flies than a dry one. Well hidden in a forest of wild flowers and grasses, a corncrake was rasping.

We found yellow water-avens there, and some pink purslane.

Derelict cottages were overgrown with ivy which thrived thickly inside and out of the walls.

Marsh thistles and ragged robin are now in full bloom.

JULY

~ A summer-petalled sun
　　　　With penetrating rays
Commands the green earth run
　　　　Her course of seasons, days
And shadow-latticed hours
　　　　Providing subtle ways
For effervescing flowers ~

Garth Waite

~ The Latin names for July flowers
on Easdale Isle are in the index. ~
. . . . .
A. SLENDER ST. JOHNSWORT
B. FUCHSIA
C. SORREL　　　c. and e · · · · SMALL COPPER
D. KNOTTED PEARLWORT　　　　BUTTERFLIES
E. MOUSE-EAR HAWKWEED　　and LARVA
F. FIELD FORGET-ME-NOT
G. EYEBRIGHT
H. BROAD LEAVED WILLOWHERB
I. RED CLOVER　　J. GREY FIELD SPEEDWELL
　　　　K. SPEAR THISTLE　k. LACE WING

~ COMMON SHREW ~

July 5

    On our way to the bridge
we spied a tiny shrew crossing the road in great haste.
    His sharp nose and vel-
vety coat identified him.
    One cannot imagine a
shrew sauntering, as the
voracity of his appetite re-
quires that he rushes from
one meal to the next.
    The shoreline at the bridge,
was decorated with mauve
and yellow flowers of the sea
aster.

July 7

    So many flowers are coming into bloom, it would
be a marathon task to record all in pencil and paint.
We discovered delicate fragrant orchids, red broom-
rape among the thyme, and wild roses blooming.
The evening was sunny, and, walking the island, we saw many common blue butterflies
which were enjoying the warmth. Towards sunset we found a pair roosting on a stem,
the male with outspread wings, head downwards
and the female close to him with folded wings.

July 10

    Waiting for the ferry we spotted a
couple of lion's mane jelly fish near
the quay. They were brought to the
quay by the incoming tide, but were
also propelling themselves with
rhythmic contractions. Their blood
red centres were conspicuous, and
their white tentacles menacing.
Back on the island, went to see
Rosie's two male kids, now
three weeks old. Rosie was
tethered, but Annabella was
frisking happily with them.

A. SEA PLANTAIN ~	K. PINK YARROW
B. BELL HEATHER ~ b. TWO SPOT LADYBIRD	L. BUGLE l. GRASSHOPPER
C. SELF HEAL ~ in flight	M. LADY'S BEDSTRAW
D. WOOD SAGE	N. CUT-LEAVED CRANESBILL
E. RED VALERIAN e. COMMON BLUE	O. SEA CAMPION o. CINNABAR MOTH
F. GREATER BINDWEED STEM BUTTERFLIES	P. WILD THYME p. LOOPER CATER-PILLAR
G. BIRD'S FOOT TREFOIL g. LARVA COMMON BLUE	Q. ORANGE HAWKWEED
H. COLUMBINE (WILD) h. HOVER FLY	R. MEADOW VETCHLING
I. ENGLISH STONECROP i. ANT	S. FIELD SPEEDWELL
J. HERB ROBERT j. GARDEN SPIDER	T. CREEPING BUTTERCUP

SUMMER CROWD

July 13

Returning from a visit to Inverness, we spent a night in the caravan at Bogroy before driving home through Beauly and Drumnadrochit. After a drizzly start the weather improved and we crossed the moors near Foxhole; made a brief stop and walked among cross-leaved heath and bell heather. There were several specimens of butterwort. This plant ingests the nutrients provided by insects that become trapped in its sticky secretions exuded by its leaves. It will also feed upon heather debris that falls upon it
We heard a grouse calling in the distance and watched a hen harrier quartering the moor. At Loch Ness the sun was shining over the shade of oaks and silver

Speckled Wood
Twice natural size. (Pencil

birches. We watched several leaf warblers in the canopy above while a speckled wood butterfly alighted on a stone at our feet.
Throughout the journey we noted the luxuriance of the ferns and bracken. Against this green backcloth there shone out a profusion of wild roses in pale and deeper pink while the roadside was decorated with masses of red campion. Yellow irises bloomed in the wetter places interspersed with snow white cotton grass. We saw an emperor dragonfly emerge fresh from his watery nursery. Everywhere there was an abundance of foxgloves and nearer home we were delighted to come upon a field containing pale mauve and pure white varieties, and near Loch Feochan found globe flowers for the first time.

July 15 - St. Swithin's Day - Weather fine.

July 21
A few steps away from the cottage a meadow-brown butterfly alighted with outstretched wings in front of us. Harebells and blackberry blooming, and bell heather making brilliant display. Lovely rainbow to-day, this time over Ellenabeich.

July 13

Cotton grass. - Eriophorum angustifolium.
& Globe flower. (Trollius europaeus)

Butterwort. (Pinguicula vulgaris) Top right.

Looking into the head of globe flower.

Pencil study of
Head of Buzzard.
Yellow eyes. nut-brown iris
Wing span 120-125 cm.

24/july
In caravan.

July 25

July 24
A section of the abandoned slate quarries
on the isle of Luing forms three vertical faces
of a cube open to the sky. Entering it was
something like standing in a cathedral.
Looking up we saw a buzzard making an un-
hurried and silent passage from one pinnacle
to another. The underside of its outstretched
wings was clearly visible.
The weather was cooler than July 4th when
I took my first swim.

On Seil we found some ladybird's eggs attached to the underside of a leaf. We brought
them home and now they are hatching. I took some specimens of the emerging larvae for
examination under a microscope. At a magnification of fifty times it was easy to con-
firm their identity. The black head and strong jaws, the six jointed legs, the grey seg-
mented body with black spots and a few hairs, and notably the sucker pad on
the tail which they used immediately for stability.
I divided the family into two communities, and distributed them on two rose bushes.
After twenty four hours they had dispersed into several small groups. From the look of
the rose leaves on which they rested it appeared that their first meal might have been
a shallow scrape of the leaf surface. They did not appear to eat their own egg shells.

SUMMER ON SEIL

~ V Waite ~

FOR FLOWER NAMES SEE END OF INDEX

~ EASDALE ISLAND~

WHITE CLOVER ~ Think of me.
FOXGLOVE ~ Insincerity. SAINT JOHN'S WORT ~ Animosity.
FUCHSIA ~ Taste. SORREL ~ Affection. HAWKWEED ~ Quick sighted
FORGET-ME-NOT ~ True Love. WILLOW-HERB ~ Pretension. RED CLOVER ~ Indus
try. SPEEDWELL ~ Female fidelity. THISTLE ~ Retaliation.
HEATHER ~ Solitude. WOOD SAGE ~ Domestic virtue. VALERIAN ~ An accom modating
disposition. TREFOIL ~ Revenge. COLUMBINE ~ Resolution. STONECROP ~ Tranquillity
CRANESBILL ~ Steadfast piety THYME ~ Activity. BUTTERCUP ~ childishness.

FLOWER PLATE Page 80 ~ SEIL ISLAND ~ FLOWERS ON THIS PAGE.

POPPY ~ Silence. GERMANDER   SEA ASTER ~ Afterthought.
    SPEEDWELL ~ Facility.   PINK WATER-AVENS.
    OX-EYE DAISY ~ A token.   SKULLCAP. (blue)
    For grasses ~ see index.   LOUSEWORT. also CRANE FLY
    IRIS ~ Message. P. 82   on PYRENEAN LILY ~ Gaiety.

FOR JULY ~

MEANINGS

FLOWER

To-day I saw the dragonfly
  Come from the wells where he did lie
An inner impulse rent the veil
Of his old husk from head to tail
Came out clear plates of sapphire mail
He dried his wings like gauze they grew
Thro' crofts and pastures wet with dew
A living flash of light he flew

(Unidentified Poet)

DRAGONFLY on
    YELLOW IRIS . . .
      Iris pseudacorus.

Jul 26

On the ridge today we came across a small caterpillar in smart yellow and black uniform which will eventually turn into a six spot Burnet moth and a little later on we saw such a moth on some tufted vetch.
Overhead a meadow-pipit with a beak stuffed with insects was almost hysterical with alarm lest we should discover her well hidden brood.
The honeysuckle is in bloom, marsh angelica, wild carrot, and large headed harebells in deep blue and white thrive in the most unexpected places.
Heath speedwell and eyebright are dotted among the grasses, the latter being used by herbalists in the middle ages to prepare a powder for brightening the eyes.

. . . Burnet moth
and caterpillar
on tufted vetch

. . . The ridge is
the best vantage point
for viewing the dis-
tant islands. Today
Colonsay shone in
the Summer sunshine

. . . .

West of Jura, south of Mull,
North of Islay herring gull
Tern and gannet skim the blue
Where they glide I would go too!

Where the geese so often fly,
Forming skeins against the sky
Where the seal and shark and whale
Wrestle with the swell and gale.

In the great Atlantic way
Colonsay and Oronsay,
Inner Hebridean twins
Part the sea like porpoise fins

With Columba as my guide
I would travel with the tide,
Visit these enchanting isles
Gladly counting the sea miles

A distant prospect of Colonsay . . . G WAITE

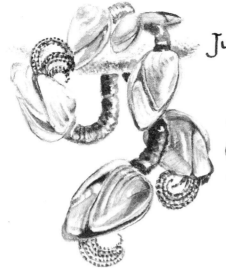

Jul 27

On rough days the shore becomes littered with the most unlikely objects thrown up by the waves and left by the receding tide. We wondered who could be the owners of a lady's high heeled shoe, an unopened carton of milk, a leather punch ball, and a gold fish. More common are pieces of drift wood (of which our mantle-shelf is made) mooring buoys, a plastic fender. The latest addition to our collection of mooring-buoys was a metal sphere encrusted with goose barnacles. As recently as the seventeenth century it was believed that each barnacle housed a tiny bird whose legs were visible protruding from the shell. This little bird was thought to be a goose — hence the name
Barnacle goose...

GOOSE BARNACLES

BLUE JELLY FISH . . . . .
Half the size of those usually seen.

Jul 28

Walking the shore I found several Blue Jelly fish stranded at the high water mark. The four horseshoe-shaped reproduction organs were a lovely amethyst colour. This creature is also called ~ Moon jelly ~

Jul 30

The Lion's mane jelly fish that we saw near the quay earlier on is rarer and will grow to two feet across. The Blue jelly fish is usually three to eight inches, but may be more.

This morning my attention was attracted to a bird fluttering among the rocks. Moving closer I discovered it to be an Arctic Tern obviously in distress. There was no mark of external injury and I supposed it had succombed to some parasitic infection. Holding it in the hand one could see the beauty of its blood red beak and small webbed feet, but its dark eyes wore a glazed expression. To have left it a prey to cats, dogs or gulls would have been less kind than a decision to dispatch it quickly by a sharp blow on the head.

... LION'S MANE JELLYFISH
Approx. one quarter natural size

AUGUST

THE BAY
FROM THE OLD
COAL BUNKER

Fishermen go out from here in search of Nephrops (Dublin bay prawns).

August

Ceud mios an Fhoghair   The first month of Autumn.

The pleasant'st angling
is to see the fish
Cut with her golden oars
the silver stream.

Shakespeare

Aug 1
After the July flower explosion
most of the specimens are still
blooming with some additions. We
discovered tutson ~ well berried,
knapweed, golden rod, ragwort in
profusion, common rush and montbretia. A male merlin gave chase
to a pipit over one of the quarries.

. . . . . .

Aug. 2
Walking up the forest path at the side of Loch Seil. we met a fisherman
returning from a smaller loch higher up. He told us he had permission
to take the trout there. In the damp grass we saw a golden coloured
moth, which though not positively identified, we believed to be an early
thorn moth. The clear water from which
our supplies are drawn was
dotted with white water-
lilies. Above the stems
meadow sweet, and
valerian, several
demoiselle dra-
gon flies were
in the
air.   .

WATER SPEEDWELL (dark blue)
    Veronica anagallis-aquatica
RAGGED ROBIN — Lychnis flos-cuculi
WATER FORGET-ME-NOT
              Myosotis scorpioides

      BANK VOLE

SICKENER TOADSTOOL (Top)
    Russula — emetica
HARD-FERN   Blechnum spicantux
CROSS-LEAVED HEATH. Erica tetralix

      CHIMNEY SWEEPER MOTH

TOADSTOOL
    Clitocybe
CREEPING THISTLE
    Cursium arvense

Aug 4.

In Glen Lonan, halfway between Taynuilt and Kilmore, we came to a patch of ground thickly overgrown with creeping thistle, and flanked by hazel and birch. Among the meadowsweet, ragged robin and water forget-me-nots, a small pearl-bordered fritillary butterfly sat with outspread wings, and in the undergrowth among the debris were several specimens of toadstool, hard to identify, but probably clitocybe.

On a sprig of cross-leaved heather among a variety of stems and leaves, a a chimney sweeper moth was resting, its sooty habit relieved only by the tiny white extremities of its forewings.

In many places the turf was criss-crossed with vole runs. Repeated journeys to and fro by these little rodents had formed grass tunnels, and more than once we caught sight of the tail end of one of these travellers disappearing into its well worn covered track.

On the moors the fat lambs so quickly grown were lying among reed clumps in an attempt to escape pestering flies.

Many birds have finished nesting and have become quiet, but the gurgle of the burn broke the otherwise complete silence. The carcase of a dead crow lay close to the turf, and several sexton beetles in their black and orange livery were already busy with the burial.

On our way home, we picked some fine specimens of purple loosestrife, Himalayan balsam, and found woundwort and sneezewort in flower.

Aug 5

Early Autumn mists are already obscuring the brightness of the Summer sunshine creating the characteristic haziness of the Hebridean scene. Herons have been absent since the breeding season, but we observed one lonely individual flying towards us out of the mists of Mull. It saw us first and banked away towards Insh.

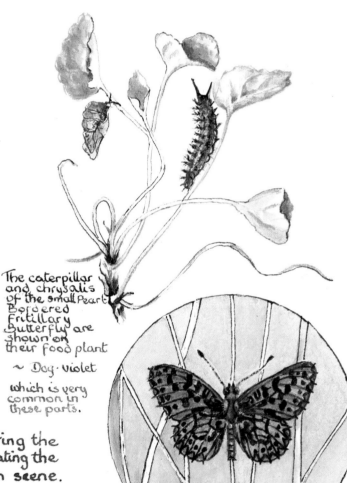

The caterpillar and chrysalis of the small Pearl Bordered Fritillary butterfly are shown on their food plant

~ Dog-violet

which is very common in these parts.

SMALL PEARL-BORDERED FRITILLARY

VICKY WAITE

Aug 6

Although the day was warm, a large toad with one blind eye forsook a damp cushion of saxifrage and remained exposed on the garden path for several minutes.

Aug 7

A blue-tit visited the garden this morning, twittering to himself and feeding alone. He was no doubt enjoying a day's excursion from Seil. Old gnarled fuchsia bushes defy the elements among the rocks in exposed places on the island. Some of them enjoy the protection of the slate-walled grottos, the ruined remains of the quarrymen's activities a century ago. Soon the same habitat will afford a lodging for the montbretia which grows in wild profusion here.

Aug 9

A morning walk at 7 o'clock, weather clear and fresh, tide high, wind northerly. A flock of about a hundred starlings flew urgently to begin their day. I climbed the ridge, and with the field-glasses watched two cormorants flying very close to the calm sea before they gently splashed down. They did not dive while I watched, but seemed to adopt a mood of incredulity at the presence of each other. A wheatear left the path ahead as I descended.

Aug 10

Watching a hoverfly I compared its ability to remain stationary in the air, or to fly with that of a helicopter. Seeing it alight upon dusty ground. I observed another characteristic shared with that machine, for a little cloud of dust was raised by the air currents from its tiny wings.

· · · · ·

Aug 12

A very sunny morning followed a wet night. The stems of many plants including honeysuckle now carry blobs of cuckoo-spit. Inside the bubble mass hides a bright green frog-hopper which sucks plant juices. Later in the Summer it will turn brown, its wing-cases will harden, and it will leap and fly in joyful abandon, having forsaken its frothy lair.

~ All sorts of tiny toadstools are springing up on the island ~

HONEYSUCKLE
WITH
CUCKOO SPIT

Aug 13

The grouse shooting usually starts on August 12ᵗʰ. The glorious twelfth is celebrated a day late this year because it falls on a Sunday. We watched I A's cat stalk a butterfly this morning. To our surprise it digested it with relish. In the warmth of the evening we made our way to the northern tip of the island which looks across the Firth of Lorn. At almost every step we flushed meadow-browns from the grass. A trawler was making her way back to Oban fish dock. It was smothered with a white cloud of gulls. Fishing from the boat near the derelict pier, I took up a young angler fish. The hideous face of this voracious creature was relieved by the beautiful orange brown and cream mottled colours of its underside.

COMMON MUSSEL SHELL encrusted with BARNACLES.

Spirorbis borealis

Single tube enlarged

BARNACLES ... Balanus balanoides

viewed from above.

Aug 14

At breakfast time I looked through the back window, and could see a redstart among the lilies. As I watched this beautiful bird, he fluttered across to the kitchen garden and pecked among the onions and flowering runner beans.

Visiting the water meadows between sea and wood at Ardfad, we picked up some amusing mussel shells encrusted with barnacles. They looked like the plumed hats of ambassadors. Many empty shells lay about there, fragments left by oyster catchers. Among a forest of marsh grasses numerous scotch argus butterflies were synchronizing their flight and their hiding deep in the grass, with the alternation of sunshine and shower as the clouds passed over. On a stone, warmed by the sun among bell heather, sat a grayling butterfly. Its folded wings were so well camouflaged as to make it scarcely visible.

Back home, among the thicket a pair of linnets, the cock with gorgeous crimson feathered breast, have been chirruping prettily as their fledged young disported themselves in the evening sunshine. These lively and tuneful little finches have been favourite cage birds frequently hybridised with canaries. The resulting mules are said to be fine songsters. Found some brilliant flame coloured anemones in a rock pool. We were reminded that at one time a profusion of lovely coloured anemones clung to the wooden pier. We wondered what disturbed them. Two pollocks were tied to our ship's bell outside the cottage door this evening. No doubt a present from a neighbour. Had a good fish kedgeree for supper.

Aug 15

We climbed the ascent overlooking Loch Seil by the forestry path, stopping to pick the odd ripe raspberry. A brown dragonfly flashed past grass of parnassus and a dark green fritillary butterfly followed the tiny burn fringed with water forget-me-nots, which ran beside the track. A sweet amber tutsan was growing on the bank, and higher up, we disturbed a hedgehog snuffling about in the undergrowth. He curled up rapidly on our approach. On our return we saw a common lizard crossing the path. I managed to catch her gently. She was about 12 cms. long obviously gravid and near her time. Having found only one lizard on Easdale Isle we immediately decided to increase the stock by introducing her. She travelled the short journey home installed in a little nest of grass tucked into a screw-topped jar. I liberated her on the sunny shelf of the ridge facing west. She lay enjoying the warmth for a while before moving away inside the cover of grass and heather. Her name quickly became Elizabeth, and we shall wonder how she is faring.

Vicky was particularly interested in the yellow and black ichneuman flies near the forestry path. They were hunting for caterpillars in which to lay their eggs. At home we congratulated ourselves on having driven out of the forestry area just before an official arrived with padlock and chain to secure the gate.

Aug 19

A painted lady butterfly alighted delicately on the flowering head of a spear thistle. This plant with its weapon-like leaves and menacing prickles seems to be the thistle adopted by Scotland as its national emblem.

We feel the spirit of Scotland is so well expressed in the lines of Sir Walter Scott from The Lay of the Last Minstrel, 'Land of the Mountain and the Flood' being taken as the title of Hamish Mc Cunn's orchestral piece. This lovely music with its poignant passion and patriotic fervour is a much loved favourite of ours.

~ A rather torpid garden-tiger moth came to settle in the lily tub this morning, later we found him resting on some blackberry about to flower.~

~ Land of brown heath
       ~ and shaggy wood
Land of the mountain
       ~ and the flood
Land of my sires
       ~ what mortal hand
Can e'er untie the filial band
That knits me
       ~ to the rugged strand.~

ICHNEUMON FLY
PAINTED LADY BUTTERFLY
(showing underwings)
RUBY-TAIL WASP
GARDEN TIGER MOTH
MASON WASP

SPEAR THISTLE (SCOTCH THISTLE)
Cirsuim vulgare and BRAMBLE

PAINTED LADY
BUTTERFLY . . . .

~FLOWERING ON SEIL~

Easdale Isle and Mull in the distance.

COMFREY
Symphytum
officinale

WILD ROSE HIPS...

MEADOWSWEET . . . .
Filipendula ulmaria

ROSEBAY WILLOWHERB. . . .
Epilobium angustifolium

. . .WILD ROSES (unusually) late flowering

Aug 20th
A. Male Merlin gave
chase to a rock pipit across
one of the quarries

~ FIELD NOTE BOOK ~

Slate blue
upper parts — Black
banded tail — Under
parts rufus striped on
buff

Meadow Brown
Caterpillar

~August~
Easdale Island

Wild Strawberry
Fragaria Vesca

Meadow Brown
Butterfly

Lesser   Spearwort
Ranunculus
flammula

Scottish Bluebell
Campanula rotundifolia
Bell Heather
Erica cinerea
Wild Carrot
Daucus carata

GREAT BINDWEED Calystegia sylvatica

Aug 23

The slated roof of a shed near us has attracted leaf-cutter bees which nest in the crevices. With scissor-like jaws, the bee alights on a selected leaf, and cuts shaped sections to line her nest. Rose leaves are commonly used, but one bee visiting our garden preferred the leaves of a false acacia. We were impressed by the tremendous haste and delicacy with which these insects set about their task.

LEAF CUTTER BEE

Aug 29

A female garden spider is at the centre of her beautifully constructed web, one of its guy ropes being attached to a rye-grass stem. The tension has pulled the stem.

into a curve more than 20° out of perpendicular. The dark coloured spider has a striking white cruciform pattern reminiscent of a fleur-de-lys.

MINING BEES on the sun-baked slate flag-stones.

Aug 31

In the shade of the fuchsia tree in our garden, we have placed a dish of water. Blossom our tame blackbird comes to bathe here. The rowan berries are ripening now, and she has already begun to feed on them. Mining bees are mating frantically A common whelk shell has appeared on the mantle-shelf. It is smooth and white, no doubt scrubbed and tended by my industrious wife. It measures nearly six inches in length.

Carder Bee, see page 97.
'Flower Meanings' for August are on page 109.

SEPTEMBER

FLOODED QUARRY
BEHIND THE CAFÉ
AT ELLENABEICH
SEIL

September

Mios meadhonach an fhoghair, an seachd mios
~ Middle of Autumn.

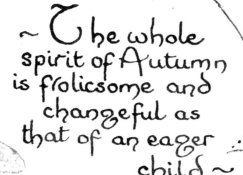

~ The whole
spirit of Autumn
is frolicsome and
changeful as
that of an eager
child ~

From an Essay
by Roger Wray

The delicate flowering
HIMALAYAN
KNOTWEED

Polygonum
polystachyum
makes a
pink haze on
the banks of
some of the
burns on Seil

Sept 1

The normally shy otter which evaded us on
our early morning vigil in June, gave us a
delightful surprise at mid-day to-day.
Our visiting cousins were with us on Clachan
Bridge when through binoculars we watched
an otter playing in shallow water. Each time
it dived its tail broke the surface. Local resi-
dents say that otters are frequently seen there.
At Ellenabeich, the baker, up and about at
an early hour watched an otter bring her
cubs into the quarry behind the café, but
such sightings as these are rare.

Sept 2

When, a year ago, Rosie the goat produced
twins, we delighted in watching the trio on and
about the ridge. This year's pair were both
billies which we miss now that they have
been sent away.

DAMSEL FLY on

GRASS of PARNASSUS
Parnassia palustris

Female Otter

~Autumn comes,
        and with her snare
Catches trees all unaware,
Paints their leaves, then makes them bare
    ~ With merry glee ~

Then the days much shorter grow
Stronger, cooler winds do blow,
Swallows o'er the sea must go
    ~ To warmer lands ~

Ragwort ~ Birch ~ Redstart ~

All the sky doth
    turn dark grey
As she passes on her way;
Duller then gets every day
~ Till winter comes ~

~ G Waite ~
(When a schoolboy)

~Bramble~ ~ Marsh Angelica ~ ~ Seeding Dandelions~

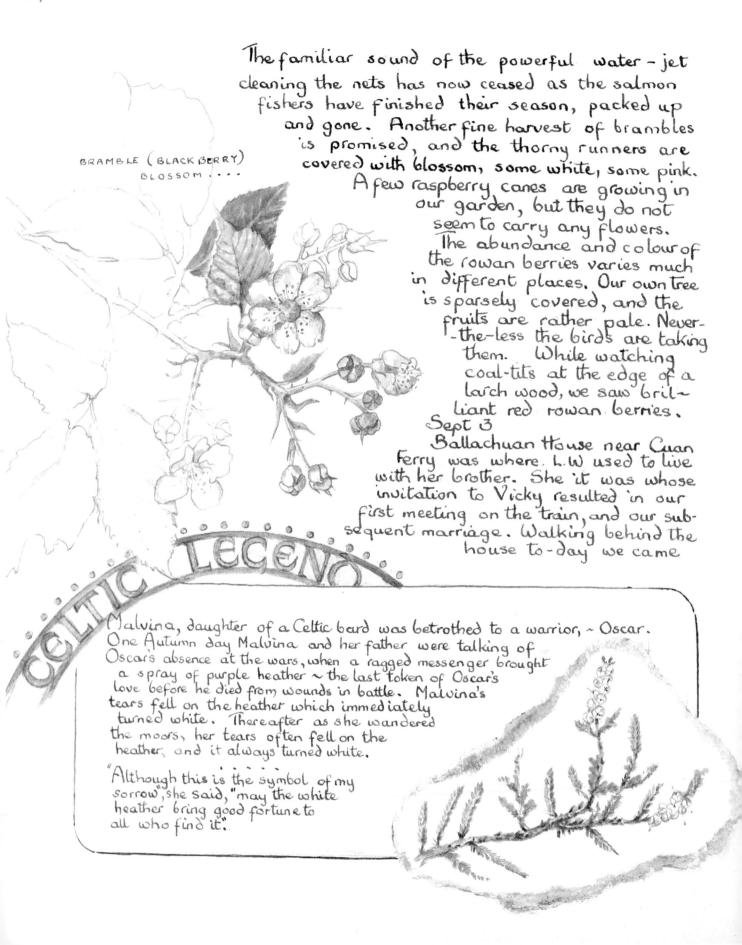

The familiar sound of the powerful water-jet cleaning the nets has now ceased as the salmon fishers have finished their season, packed up and gone. Another fine harvest of brambles is promised, and the thorny runners are covered with blossom, some white, some pink. A few raspberry canes are growing in our garden, but they do not seem to carry any flowers.

The abundance and colour of the rowan berries varies much in different places. Our own tree is sparsely covered, and the fruits are rather pale. Nevertheless the birds are taking them. While watching coal-tits at the edge of a larch wood, we saw brilliant red rowan berries.

Sept 3

Ballachuan House near Cuan Ferry was where L.W used to live with her brother. She it was whose invitation to Vicky resulted in our first meeting on the train, and our subsequent marriage. Walking behind the house to-day we came

BRAMBLE (BLACKBERRY) BLOSSOM....

CELTIC LEGEND

Malvina, daughter of a Celtic bard was betrothed to a warrior, ~ Oscar. One Autumn day Malvina and her father were talking of Oscar's absence at the wars, when a ragged messenger brought a spray of purple heather ~ the last token of Oscar's love before he died from wounds in battle. Malvina's tears fell on the heather which immediately turned white. Thereafter as she wandered the moors, her tears often fell on the heather, and it always turned white.

"Although this is the symbol of my sorrow", she said, "may the white heather bring good fortune to all who find it".

~ SQUAT LOBSTER ~

This creature is described
as widely distributed, but
not common.
Fishermen tell us that squat
lobsters occasionally enter
their creels put down to
catch prawns and baited
with dead fish.

Fierce and uncommon denizen
of the lower shore (with tail
turned under)

Natural
Size

FIELD NOTE BOOK SEPT 3rd 83

across the lovely caterpillar of an Emperor moth feeding on
meadowsweet. We brought it home and hope to rear it, and watch
the emergence of the moth. Himalayan Balsam and Himalayan
Knotweed both grow in the area. At the ferry we watched two spiny
spider crabs in a mystical embrace among the weeds in shallow
water, and turning over a stone we experienced our first en-
counter with a squat lobster.

Sept 10
The emperor moth caterpillar which we have named Septimus (because
we found him in September) has settled down in an up-ended plastic
bag. Given the choice of meadowsweet, bramble and raspberry
leaves, he has chosen the latter. A leafy twig stands in a bottle of
water. He does nothing but eat, and judging by the copious droppings
he continues through out the night. We begin to wonder when his
internal mechanism will give him the signal to commence pupation
This morning I feared we had lost him, but after searching the room

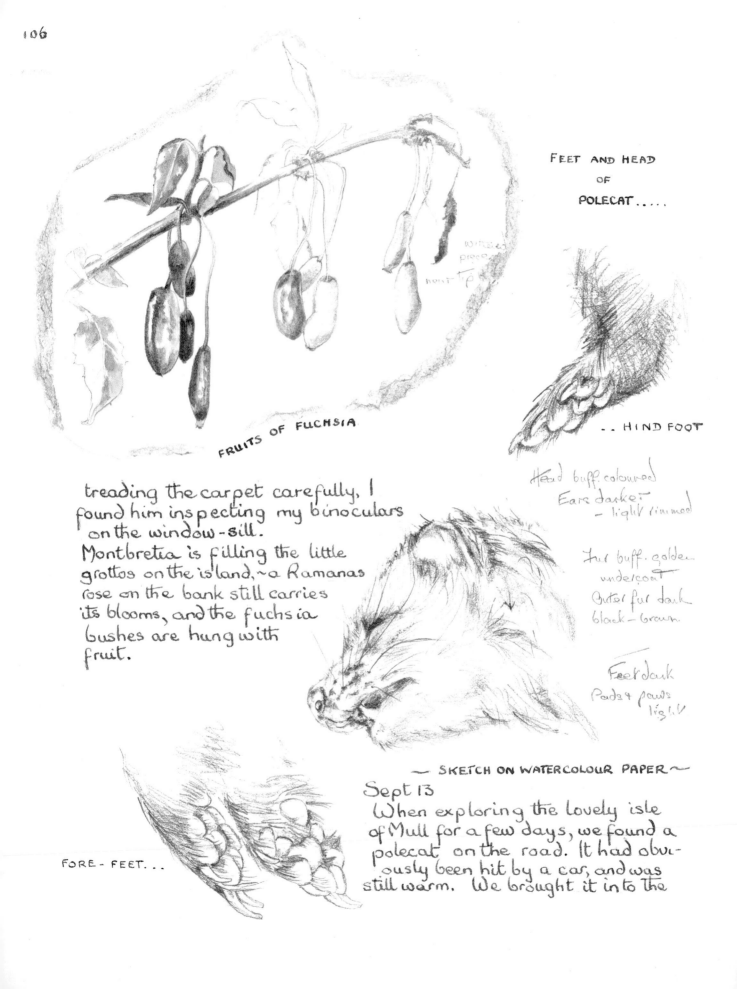

FEET AND HEAD
OF
POLECAT.....

FRUITS OF FUCHSIA

.. HIND FOOT

Head buff. coloured
Ears darker
- light rimmed

Fur buff. golden
undercoat
Outer fur dark
black - brown.

Feet dark
Pads & paws
light

treading the carpet carefully, I
found him inspecting my binoculars
on the window-sill.
Montbretia is filling the little
grottos on the island, ~ a Ramanas
rose on the bank still carries
its blooms, and the fuchsia
bushes are hung with
fruit.

— SKETCH ON WATERCOLOUR PAPER ~

FORE-FEET...

Sept 13
When exploring the lovely isle
of Mull for a few days, we found a
polecat on the road. It had obviously been hit by a car, and was
still warm. We brought it into the

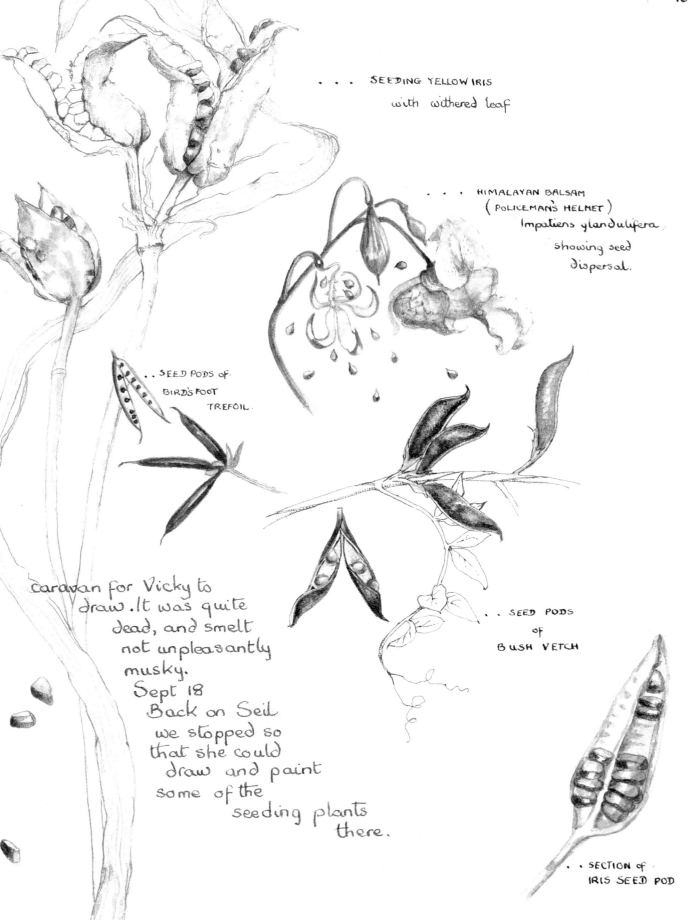

. . . SEEDING YELLOW IRIS
with withered leaf

. . . HIMALAYAN BALSAM
(POLICEMAN'S HELMET)
Impatiens glandulifera,
showing seed
dispersal.

. . SEED PODS of.
BIRD'S FOOT
TREFOIL.

caravan for Vicky to
draw. It was quite
dead, and smelt
not unpleasantly
musky.
Sept 18
Back on Seil
we stopped so
that she could
draw and paint
some of the
seeding plants
there.

. . SEED PODS
of
BUSH VETCH

. . SECTION of .
IRIS SEED POD

AUTUMN ON EASDALE

—V.Walker

FOR FLOWER NAMES SEE END OF INDEX

There flew into our room at lunch-time to-day a wasp, blacker and more slender than the common social variety. It proved to be a specimen of the genus brabo. A solitary wasp, it specializes in digging a tunnel in the pith of plant stems, or in the ground in which it lays its eggs. Its provision for the grubs is usually flies which are paralysed by its sting being stored with the eggs. We saw the first peacock butterfly and ate our first blackberry picked from the tip of a bramble.

. . . HAZEL NUTS

Flower~Meanings for August.

Water Lily ~ Purity of heart. Creeping Thistle ~ Austerity. Water-speedwell ~ Female fidelity. Ragged Robin ~ Wit. Water-Forget-me-not ~ True Love. Hard-Fern ~ Sincerity. Cross-leaved Heath ~ Solitude. Fuchsia ~ Taste. Honeysuckle ~ Sweetness of disposition. Wild rose ~ Pleasure and pain. Meadowsweet ~ Uselessness. Rose-bay Willowherb ~ Pretention. Scotch Thistle ~ Retaliation. Lesser Spearwort ~ Ingratitude. Scottish Bluebell (Harebell) ~ Submission. Bell Heather ~ Solitude. Greater Bindweed ~ Insinuation. Acacia ~ Chaste Love.

~ To Autumn ~

Season of mists and mellow fruitfulness,
Close bosom-friend of the maturing sun;
Conspiring with him how to load and bless
With fruit the vines that round the thatch-eves
run
To bend with apples the moss'd cottage-trees
And fill all fruit with ripeness to the core;
To swell the gourd,
and plump the hazel shells
With a sweet kernel; to set
budding more.......

John Keats

Coal-Tits on Rowan.
Pencil drawing.

~ FLOWER MEANINGS ~

Bramble ~ Lowliness. Birch ~ Meekness. Marsh Angelica ~ Inspiration.
Dandelion ~ Oracle. Himalayan Balsom ~ Impatience. Hazel
~ Reconciliation. Hawthorn ~ Hope. Devil's bit Scabious ~ Unfortunate
love. Field Gentian ~ You are unjust.

~Sonnet~

G Waite

When summer sun no longer burns the sward
And gossamer adorns the fading leaf
The spider that has fashioned it will board
Her silken parachute and, in a brief

Excursion, trav'lling, wind-borne, unafraid
To unknown destination take her flight.
In years of Autumn shall I so persuade
My earth-born spirit to forsake the sight

Of all I know? O grant me, Blessed Lord,
The faith to keep my grip fast hold on Thee
Believing that my journey hence, aboard
The gossamer I've spun, will carry me

From this fair summer to a fairer still
Where Thou art Sun and Season,
Way and Will.

HONEYSUCKLE

The fineness of a spider's thread is such that not much more
than a 100 grams of it would be long enough to encircle
the globe. Its lightness and strength combine to make it an
airborne vehicle which the gossamer spider uses to carry it
from one place to another.
Darwin observed the arrival of such aeronauts on the
Beagle when he was sailing nearly sixty miles from the
nearest land.

EMPEROR MOTH
CATERPILLAR

Sept 25

Our Emperor Moth caterpillar, Septimus, has now spun a silk cocoon within which the transformation will proceed. For six or seven months, the long sleep of preparation will ensue. The plastic container has been stored in the garden shed where the normal temperature will be more natural than in the warm cottage. We must be patient in waiting for the moment of emergence, and the resolution of the moth's sex.

Sept 28

In the days before and since the Autumn Equinox we have enjoyed some lovely sunrises, and a succession of vivid rainbows. On our return from the village shop we found our neighbours S. and M. extricating one of our little cypress trees from the jaws of Annabella. Chewed and battered the victim was hastily replanted in its tub outside the door in the hope that it will revive.

~EMPEROR MOTH~
female

Sept 25. The news heard to-day is of the plight of the Shetland farmers whose cattle are short of grass in the sodden moorland. It is a problem feeding them when a wet Summer has ruined the haycrop. Something that I knew to my cost in my farming days.

OCTOBER

THE ATLANTIC
    POUNDING THE ROCKS
        AT THE SOUTH WEST

            POINT OF THE ISLAND

October

Mios deire annach au fhoghair ～ A later month of Autumn

Autumn leaves are falling in the woods on Seil.

These fleshy toads tools are abounding on Easdale Isle.

～ A song for dun October
That tints the woods wi'broon
And fills wi'pensive rustling
The wooded dells aroun' ～

HYGROCYBE
The damaged ones soon lose lustre and rot.

Oct 1

We arranged with the ferryman for an 8 a.m. crossing this morning as we are visiting Edinburgh for a couple of days. The wind was cold, and the water choppy as we stood awaiting the sound of Johnnie's familiar whistle. The lobster fisherman's yellow boat was riding at anchor, an orange buoy swinging rhythmically from its bows. Nearby a grey seal pushed its head above the water surface to snatch a breath of air before diving out of sight.

On the road to Edinburgh we saw a sparrowhawk dart across our path. We had a picnic lunch near a lime-tree on the roadside. No doubt the local bees had taken their fill from this tree. Further on we passed a Royal Mail Bus that carries passengers as well as Mail to remote farms. In colour, it seemed to link itself with the rowan berries cheering up the countryside.

..BLACK GROUSE (female)

Oct 5

We parked overnight beside Loch Earn,
its waters navy-blue flecked with white.
Above the shoreline the grassy banks abound
with trees and shrubs. In their shelter are
inkcap and fly agaric fungi among the
roots. A solitary mole-hill, recently thrown
up was among the ferns, ground elder and
wood sorrel. The ground was littered with
broken and gnawed hazelnut shells, obviously the work of mice and voles.
Before turning in we walked along the
waters edge, following a dipper that me-
thodically explored the loch bed in the shal-
lows. Overhead small swarms of may-flies
were dancing. The willows on the bank
had been attacked by sawflies whose
eggs are encapsulated in the characteristic galls decorating the leaves.

Oct 6

On our return journey we saw several kes-
trels. By-passing Oban we drove through
Glen Lonan where the contrast of rowan berries and ling heather was a sight to remember. A black grouse flew alongside for a
minute or two as we drank in the beauty
of the place.

FLY AGARIC
TOADSTOOLS

INK CAP OR LAWYER'S WIG
TOADSTOOL....

SHORT-TAILED VOLE..
(Approx. 14 cms
from nose to tail)

. . . .

The softer tones of SULPHUR TUFT . . . .
which grows on rotten tree stumps
contrast strongly with the garishness
of the HYGROCYBE.

Oct 9

On a walk round our island in the early evening, we noticed in several places a great irruption of flying ants. The signal which determines the day on which the winged pairs leave the nest to mate seems to call up all colonies in the area. Frantically the new young queens and their 'consorts', climb the grass stems and take to the air. This one marriage flight is the only occasion on which the queens will use their wings which drop off there after. Visiting the peak of the ridge we found a small group of soldier flies, rather bee-like creatures with yellow banded black bodies flying to and fro among the thistles and grasses. Several pairs were mating. Their eggs are laid on aquatic plants, and the larvae develop in pond-water feeding on tiny organisms. The adult fly is silent on the wing, but the specimen we caught buzzed loudly on being held for examination.

Oct 10

A snap decision led us to forsake the charm of goldfinches in the garden, and to drive to the Mull of Kintyre. We flushed three mallard at Balvicar. Five shags perched symetrically on a rock watched us pass Lochgilphead. The mudflats at West Tarbert were flanked by heather-cushioned slopes interspersed with bracken and birch. Twenty-five oystercatchers and a curlew paddled among the blue, green and mauve algae covered mussel, cockle and razor shells.

~ The Wee Breeze ~

The wee breeze danced in the sycamore tree
Played with the thistle-downs,
~setting them free
Startled a linnet, when taking the trouble
To clear up the seeds that were left
in the stubble
Tangled the wool which had caught on
the thorn
And frightened a cotton-tail deep in the corn
Tickled the skin of the old crofters' goose
And banged shut the door of his but and.ben hoose
In breathless confusion it sped to the glade
(Where Autumn leaves lay in the sun-dappled shade,
With one mighty scurry it shifted and sighed,
~Then it died ~

V. Waite

TUSSOCK MOTH CATTERPILLAR.

Oct 3

East of Loch Earn the hill-farming of Argyll changes to the arable of Perthshire.
Harvest over, stubble-burning is a common sight.

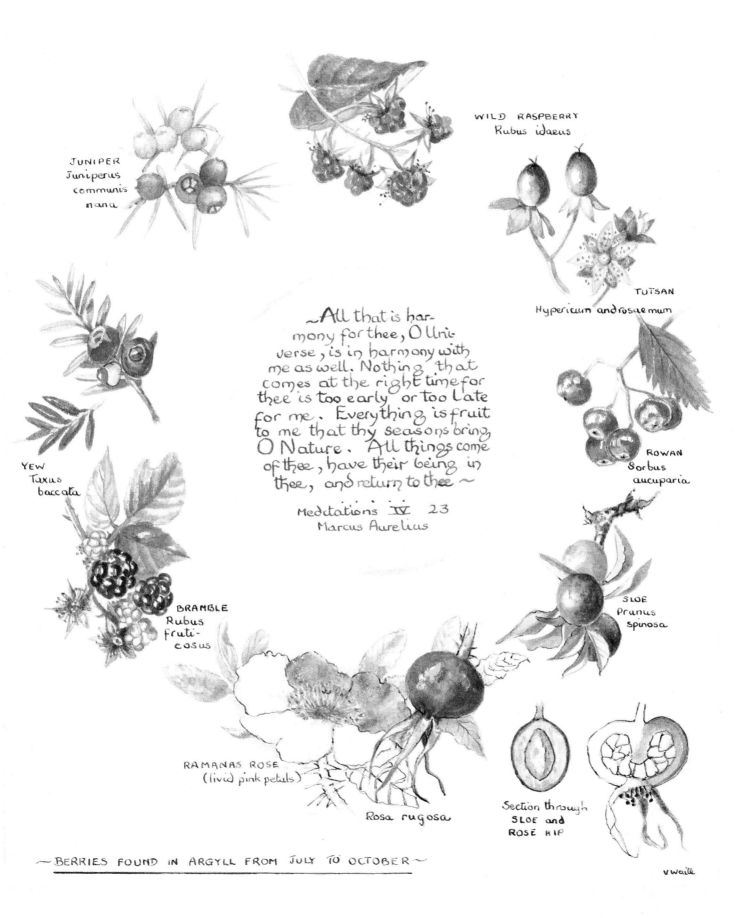

JUNIPER
Juniperus
communis
nana

WILD RASPBERRY
Rubus idaeus

TUTSAN
Hypericum androsaemum

~All that is har-
mony for thee, O Uni-
verse, is in harmony with
me as well. Nothing that
comes at the right time for
thee is too early or too late
for me. Everything is fruit
to me that thy seasons bring,
O Nature. All things come
of thee, have their being in
thee, and return to thee ~

Meditations IV 23
Marcus Aurelius

ROWAN
Sorbus
aucuparia

YEW
Taxus
baccata

SLOE
Prunus
spinosa

BRAMBLE
Rubus
fruti-
cosus

RAMANAS ROSE
(livid pink petals)

Rosa rugosa

Section through
SLOE and
ROSE HIP

~ BERRIES FOUND IN ARGYLL FROM JULY TO OCTOBER ~

V Waite

LOOKING DOWN ON
LICHENS, STONECROP, AND MOSS
GROWING ON SLATE WASTE

REDSHANK

The little bay was crammed with activity. While the seals alternately slapped flippers against their sides and adjusted their recumbent positions, the rippled sand was peopled with hooded crows, blackheaded gulls and a couple of redshanks. The flocks of feeding birds included some unlikely mixtures, starlings with pied and grey wagtail and cormorants with green plover. The cooing of eiders was an accompaniment to the staccato cry of jackdaws. On our return we saw a small flock of fieldfares, and a baby hedgehog mesmerized by our sudden approach froze in the middle of the road. As he kept so still it was easy to straddle him as we drove on. A glance in the mirror assured us he was unharmed.

Oct 29

An hour with binoculars soon passes when focusing on the weeds and scrub surrounding our garden. Parties of goldfinches included grey-pates the young of the year not yet in adult plumage Blue and great tits, as well as wrens, have investigated the nest box I put in the rowan tree in the garden. We hope for a nesting pair in the Spring.

. . . . . Seals are well camouflaged as they lie on the rocks.

Oct 30

Two new cats have joined the island's population with the arrival of their owner. We are anxious lest this should be a threat to the birds that we expect at the window-sill as the colder mornings ensue.

Among the tangle of long grass and bramble, near the spot where we had liberated the pregnant lizard We found the caterpillar of the fox moth There was still a warmth in the Autumn sunshine which it seemed to be enjoying.

GOOSE GRASS

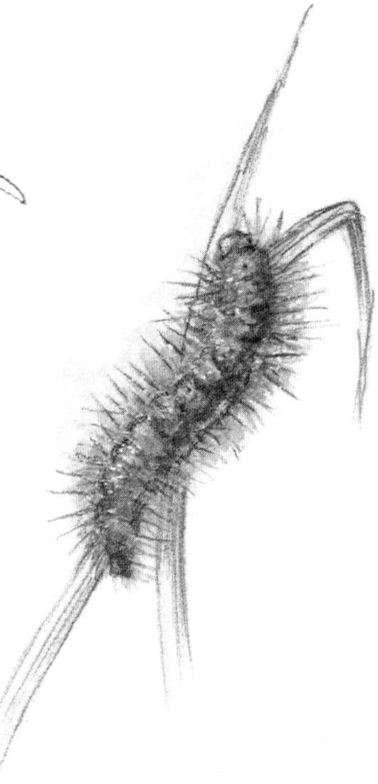

EIDER DUCK (Male)
DIPPER.
RED BREASTED MERGANSER
FOX MOTH CATERPILLAR

NOVEMBER

THE SEA EXPLODING
ON ROCKS AT
THE NORTH WEST
POINT

November

Ceud~mhios a' gheamh ~ raidh ~   The first month of goodwill

~ No shade, no shine, no butterflies
   no bees
No fruits, no flowers, no leaves
   no birds
         November ~

November. T. Hood

~ Lastly, came Winter clothed in Frize
   Chattering his teeth for cold
      that did him chill ~

The march of the Seasons.
E. Spenser

GREAT TITS

Nov 1
An early morning storm from the west awakened
us. Spectacular lightning over Mull was followed
by rumbles of thunder reverberating among the
mountains. This commotion was matched by
an angry sea, and visiting the north west
point at high tide we noticed that the crash
of the mighty waves was followed not only
by clouds of spray, but by showers of pebbles dashed against the rocks, some to the
height of three metres.

Nov 3
Great tits visit the garden more often in
Winter. We often see a solitary red~
throated diver out at sea at the southern
tip. Two domestic hens peck about in the
short grass in front of the window. Annabella delights in chasing them. To our con-
cern this morning she seized one playfully
lifting it up in the air. There was much
clucking and several stray feathers.

Scarba
from the gully,
a quiet retreat
on the Island

Nov 4

Rounding a corner at Kilninver we came in sight of a Stoat. No doubt it had seen us first as it stood upright in the middle of the road. After pausing to sniff the air it melted into the bracken, though we still had the feeling we were being watched. Several days ago a stoat had turned up near Easdale Post-Office, darting in and out of crevices in a stone wall, and watched by interested spectators.

——— · · ———

Nov 5.

In the giftshop at Easdale (Ellenabeich) Vicky found a tiny toad desperately trying to find its way out from among the tartans and oddments on the floor. Held in her hand it occupied no more space than a five pence piece.

The island children are having a bonfire on the beach to-night.

Nov 6

One of the new cats has killed two water-shrews, the only wild mammal we have yet found on the island. Their tiny bodies were left one in the rough grass and the other among the mauve and pink primroses which are brightening up the otherwise dull garden at this time of the year. We were able to examine the tiny mole-like fur, its delicate feet and claws, and to observe the keel of white hairs on the underside of the tail. In Scotland, the water-shrew is known as the water rannie.

At this time of the year when flowers are few, - we become more aware of lichens.

Vicky painted these in early October.

DEAD WATER SHREW

Nov 7

A little boy who had been beach~
combing on the island came across
a dead snake pipefish which he
brought to us. These creatures live
among seaweed to which they
anchor themselves by
their tails.

The slate debris on the sea bed is mixed with white quartzite, and studded with iron pyrites

The slate deposits and associated rocks which form most of the island are so old that there are no fossils. Tremendous pressures over millions of years have crumpled the strata into contorted shapes. Parts of the shore are littered with boulders of various sizes, which are perpetually smoothed and rounded as they roll up and down the rock apron with every tide. White quartzite is abundant in veins which permeate the rock. In one place on the island, a fold of rock encloses a deposit of that delightfully tinted quartz known as amethyst. To hack out a piece of this is very hard work indeed.

~ CRYSTALS ~

(Note.)~The bedrock of the amethyst we found was stained with metal-oxide deposits.~

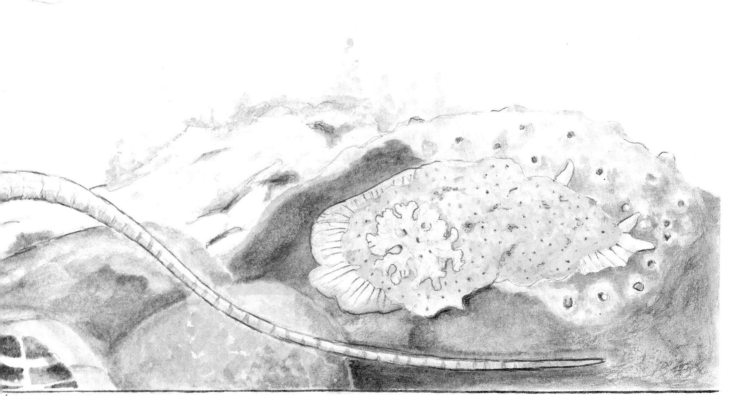

Where bread-crumb sponge occurs, it is eaten by the sea lemon, (one of the sea~slug family).

Nov. 9.

A mixed company of birds came asking for breakfast.
Blue-tits, chaffinches and a robin were there as well as Blossom and
a couple of song thrushes. Quite frequently now Charlie, our particularly adventurous blue tit pecks on the window asking for food, he is quite aggresive with the
other birds and undaunted by his smaller size.
Seil wore a new dress to-day. A severe hoar-frost clothed all the vegetation with minute
crystals turning the place into a fairy land. Some of the grasses looked particularly lovely
their heads bent gracefully with the extra weight.
Visiting the loft of our cottage in search of packing material I found that the
edges of the floor area were littered with dead bracken, the remains of a great
quantity used by earlier generations as bedding for the children who
slept there among it. A layer of this close to the eaves was
a good insulation against the winter cold.

~The other seasons, burgeoning
In lively seed and bursting bud,
Or setting fruit, have had their fling,
And now the flood
Is stilled, as if a Winter wing
Had folded, hard'ning everything
Like frozen mud.~

G. Waite

ONLY 43 DAYS
TO CHRISTMAS!

Nov 12
Vicky has sorted
her flower pressings.
The blue flowers have
kept their colour es-
pecially well, and she is
sorry she forgot to press
some harebells.

Nov. 15
A sprinkling of snow on Mull, the gales have
subsided a little. A friend on Seil introduced us
to some new territory near her home. Walking
through marshland on private property we came
to a spot where woodland skirted a little bay bath-
ed in sunshine. A small loch adjoined, and here we
saw a flock of widgeon feeding. Three mute swans
regarded the territory as their own and aggressively
opposed the attempt of a pair of whooper swans to join
them. A large roe deer bobbing its white rump as it
bounded, paused to turn and gaze at us. In several places we found fox droppings, and we were told that otters
and badgers were sometimes seen there. Vicky slipped
down a muddy bank and pulled a leg muscle, and
lost the buzzard's feather out of her hat, it was soon replaced.

Nov 23
A reluctant hibernator in the form of a small tortoise-shell
butterfly was on the wing in the garden when I went to pluck the last
of the pink roses before pruning them. The buds had been slow to bloom
there, but have quickly opened in the warmth of the cottage. They
seem to glow in the copper kettle on the living-room table.

Nov 26
Returning from Oban we passed close to a heron standing
near a burn with a small fish in its beak. The rain sodden
bracken is now russet pink and gold, and the tiny leaves
that remain on the birches look like sprinkled confetti.
Our new neighbour has left the island taking
his family and the two cats with him, ~ a
mysterious business. After a hail shower, Blossom appeared on the window-sill. We put
out some shredded suet for her.

(Nov 11th)
The Remembrance day observance was.
very Scottish. The little war memorial on
Seil was visited by a small group led
by the minister. A wreath of poppies
was laid on the hillside, while
a lone piper silhouetted against
the sky played a lament.

~FLOWER PRESSINGS~

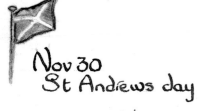

Nov 30
St Andrews day

... MUTE
SWANS

I brought home a dozen hermit crabs, all but one
in dogwhelk shells. The smallest was in a winkle.
I brought them home with seaweed and seawater and
put them in the tank. Within a few hours a small
beadlet anemone I had introduced, attached itself
to an inhabited whelk shell. The edge of its foot over-
lapping the shell mouth, making life difficult for the
crab, which soon indicated its displeasure by vaca-
ting even before finding suitable alternative accommodation.
One of the larger crabs with a tight-fitting
whelk shell quickly took the opportunity
afforded him when I put a larger shell
within reach. He examined it carefully, turning it this way and
that and thrusting his claws
into its recesses before
moving house.

The harsh metallic kronk of a
raven overhead reminded us
that the couple, paired for life
spend most of their time between
this island, and the northern tip
of Seil where they nest every year.

DECEMBER

RAVENS TO THE
EAST OF ROUGH
ISLAND

132

December

Mios meadhonach à gheamhraidh ~ Middle month of Winter

Cold December brings
the sleet

Blazing fire and
Xmas treat

The Holly and the Ivy
When they are both full grown
Of all the trees that are in the wood
The holly bears the crown.
Carol

~ Holly berries represent the
blood of Christ, and its
prickly leaves, the
crown of thorns ~

The day is done, and slowly
from the scene ~
The stooping sun upgathers his
spent shafts, and puts them
back into his golden quiver......
Longfellow

Dec 1
Celebrated my birthday.
The beautiful goldfinches have been with us all the year. The sight
of them, clothed in such brilliant colours is breathtaking, especially
on dull days.   Vicky saw a skein of greylag geese, illuminated
by the sunlight, winging their way towards Colonsay over the
ridge. Canada, barnacle, and greylags are reported to have
been seen there. Johnny's domestic geese having survived Michaelmas

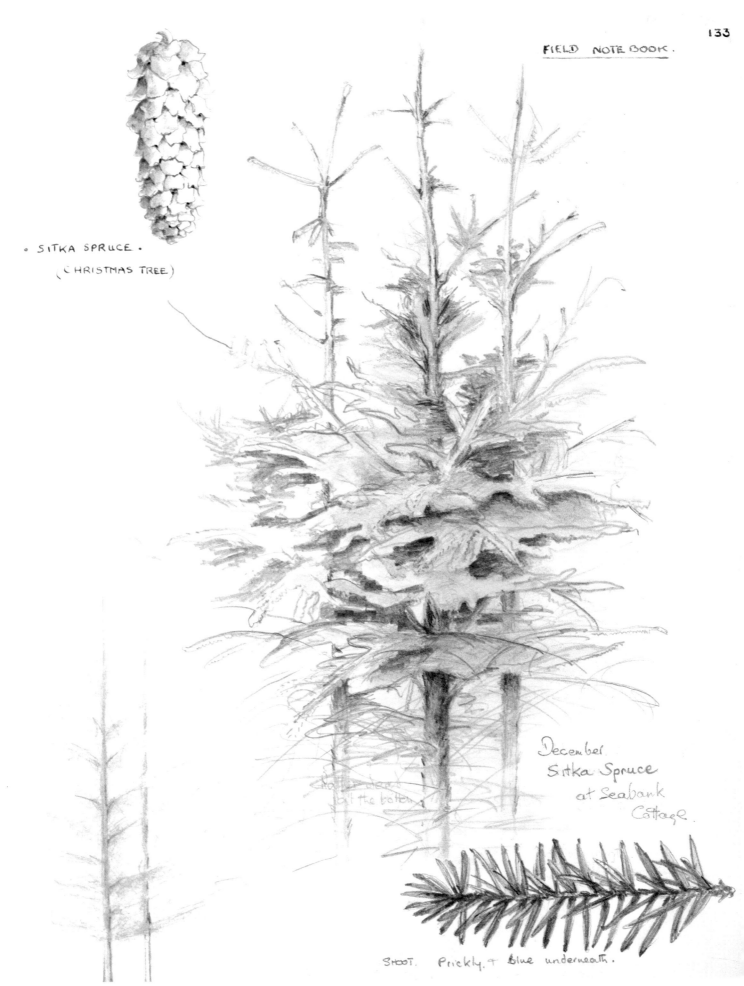

• SITKA SPRUCE •

(CHRISTMAS TREE)

December.
Sitka Spruce
at Seabank
Cottage.

SHOOT. Prickly. + blue underneath.

It is the lofty pine that by the storm
Is oftener tossed, towers fall with
heavier crash
Which higher soar . . . .
Horace

The silver birch is a dainty lady
She wears a satin gown . . . .
E. Nesbit

may be assailed by an unusual nervousness as
Christmas approaches. Sunset provided a colourful
spectacle. There was a soft pink reflective glow on the
white walls of the cottages.

Dec 4

Having few wild trees other than Rowan, Fuchsia and
Sloe on the island, we enjoy the sight of others on Seil and
elsewhere. In the area we have seen many birches, larch, sitka
spruce and a few Scots pine, as well as oak, ash, hazel, alder and
willow. Children learning their alphabet in Gaelic used to be taught
that each letter represented a tree or shrub. For example ~
B for Beith (Birch), C for Call (Hazel), L for Luis (Rowan)
S for Suil (Willow), ~ all typically Scottish trees.

Dec 7

Snowdrop bulbs planted in the garden during October are poking their

The moving finger
having writ...

moves on . . . .

shoots through the short grass. Already a number of them show a glimpse of their white blooms.

Dec 9

Tucking a book into its place in my shelves I came across Vicky's copy of Fitzgerald's translation of Omar Khayyám's poem. The words were appropriate to the wind of change which has come to the island now under the management of a new owner. We cannot help wondering how much of the island's character may change, as development plans are introduced.

Dec 10

Mandy, one of the island dogs sent Johnny's domestic geese waddling off to the water in haste. There is a Chinese goose among them and recently a stray greylag is enjoying their company.

~NOTES ON SLATE~

Exercise
Difficult.
in reverse.

- looks like fern.

Thought side of stone looked better

~Our one-eyed TOAD
is hibernating.~

SLATE STONES Dec. 1979
and MOSS.

FROM FIELD NOTE BOOK

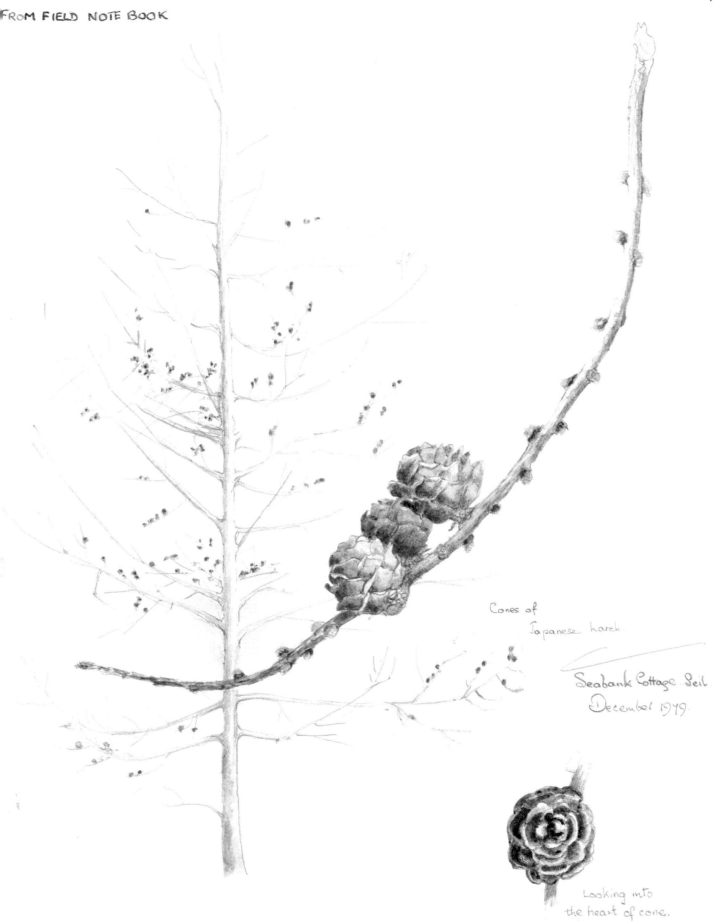

Cones of
Japanese Larch

Seabank Cottage Seil
December 1949.

Looking into
the heart of cone.

Dec 11

On an overnight journey, south of the border, we noticed large flocks of lapwings in the fields actively feeding in the light of the full moon. Returning in daylight we saw the same flocks were asleep with their heads under their wings, some standing on one leg.

Lapwings are also called Peewits and Green plover.

Dec 17

Rough weather is no stranger to North West Scotland, but it is not often that the weather forecast mentions a force twelve hurricane. Overnight the westerly gale scoured the shore, and unusually high tides pushed a mass of tangled sea weed and slate rubble twenty metres above the normal high water mark. A well worn path was obliterated, and debris littered the entrance to the donkey hollow. The individual waves had lost their identity and seemed to have clumped together into enormous walls of swell that crashed down whipping up clouds of blown spume and spray. It was difficult to keep one's feet on the ground. We wondered how the ferocity of the storm compared with that in 1881 which flooded the quarries here, and put them out of action.

~ Part of a water-filled quarry on a calm cold day. ~

Dec 18

~ A dozen slates are missing from the cottage. The channel connecting open sea with high water is blocked. Passage from quay now only possible at high water. 4 pm. Electricity failed four hours with candles ~

Dec 24 ~ Christmas Eve ~

A lovely sunrise heralded a bright cold morning, and we were encouraged to take the ferry and cross to Seil. The water was choppy, and the sight of Mull, snowcapped, and shrouded in ethereal white was once again breathtaking. On Seil there was more evidence of gale damage. The tarmac surface of the road was contorted into folds and furrows, and large boulders were distributed fifty metres inland from where they had been thrown from the shore. Several cars suffered minor damage, and one was washed from the road into the sea. A small sailing boat was holed below the water line, and lay on its side covered by the rising tide. Only its mast was visible. Among the tangle of seawrack dashed against the rocks was a fine specimen of a sunstar with twelve arms and measuring eight inches across. I brought it to the cottage window sill to dry out. Blossom came and showed her curiosity, pecking and tossing it about as she investigated.

We were amused by a report on radio news of a punch up between two Father Christmases which the police were called to sort out.

Dec 25 ~ Christmas Day ~

My son Christopher is spending Christmas with us. We talked with him about events recorded in the diary, and of much more to be learned and discovered. The Garvellachs are always beckoning, but the arrangement of a visit has so far eluded us. We also intend to get within sight and sound of the famous Corryvreckan whirlpool. The year has passed without our sighting an eagle or a whale. We talked too of friends in England. Some of them had looked askance at us when we told them of our choice of lifestyle.

ROWAN BLOSSOM in Spring, and THE ROWAN TREE in Autumn.

Dec 26 Boxing Day

At 4 a.m., an earth tremor was felt in a band fifty miles wide running from the borders in the South West to Aberdeen in the North East. The shaking was felt as far away as Belfast, and lasted for fifteen seconds. We did not feel it on the island. Friends in Glasgow at that time told us that it was a frightening experience.

Dec 31 New Year's Eve ~ Hogmanay

Going out to feed the birds, we noticed that the sun, although at its zenith, was low in the sky and casting long shadows. We watched the garden birds moving in and out of their favourite Rowan Tree. Silhouetted against the pale sky, a flock of starlings, moving as one swerved to alight. The sudden rustle of their urgent wings causing our garden friends to vanish in a flash. We turned over a few stones to find worms for Blossom, and went back to the warmth of the cottage.

New Year resolutions to remember :~

1. Bring a hedgehog from the mainland to eat up the slugs in the garden.
2. Keep some mistletoe berries until the Spring and impregnate the rowan tree with them.

~Hedgehog~

The hedgehog has no calendar
To help him count each torn-off day
Nor can he calculate how far
The burnished sun will move away
    Below horizon's rim ~

But, tightly rolled into a snore,
He slumbers all the winter through.
Oblivious of snow and hoar;
Impatience numbed. I doubt if you
    Or I could waken him.
        ~ G Waite ~

Index of illustrations

Mammals

Donkey: Equus (domestic)   36
Goat (feral): Capra hircus   22
Hedgehog: Erinaceus europaeus   140
Mouse, wood: Apodemus sylvaticus   50
Otter: Lutra lutra   101
Polecat: Mustela putorius   106
Porpoise: Phocoena phocoena   70
Seal, common: Phoca vitulina   70
  "   grey: Halichoerus grypus   12. 121
Sheep and lambs: Ovis (domestic)   44
Squirrel, red: Sciurus vulgaris   45
Shrew, common: Sorex araneus   76
  "   water: Neomys fodiens   125
Stoat: Mustela erminea   125
Vole, bank: Clethrionomys glareolus   87
  "   short-tailed: Microtus agrestis   115
  "   water: Arvicola terrestris   65

Birds

Blackbird: Turdus merula 28. 47. 55. 68. 141. (egg) 47
Bunting, reed: Emberiza schoeniculus   8
  "   snow: Plectrophenax nivalis   8
Buzzard, common: Buteo buteo   52. 79
Chaffinch: Fringilla coelebs   xIII
Cormorant: Phalacrocorax carbo   114
Crow, hooded: Corvus corone cornix   7
Cuckoo: Cuculus canorus   60
Curlew: Numenius arquata   27
Dipper: Cinclus cinclus   122
Dunnock: Prunella modularis   34
Eider: Somateria mollissima   122
Gannet: Sula bassana   57
Goldfinch: Carduelis carduelis   116
Goose, barnacle: Branta leucopsis   4
  "   grey lag: Anser anser   135
Greenfinch: Carduelis chloris   141. 24
Grouse, black: Lyrurus tetrix   115
Guillemot, black: Cepphus grylle   25

Gull, black-headed: Larus ridibundus   38
  "   common: Larus canus   14. 71
  "   great black backed: Larus marinus   33
  "   herring: Larus argentatus   72
Heron, grey: Ardea cinerea   4
Kestrel: Falco tinnunculus   6
Lapwing: Vanellus vanellus   138
Linnet: Acanthis cannabina   24
Merganser, red-breasted: Mergus serrator   122
Merlin: Falco columbarius   95
Owl, barn: Tyto alba   15
  "   tawny: Strix aluco   56
Oystercatcher: Haematopus ostralegus   40. 141
Pipit, meadow: Anthus pratensis   37
  "   rock: Anthus spinoletta   59
Plover, ringed: Charadrius hiaticula   20
Raven: Corvus corax   131
Redshank: Tringa totanus   121
Redstart: Phoenicurus phoenicurus   102
Robin: Erithacus rubecula   3
Sandpiper, purple: Calidris maritima   10
Shag: Phalacrocorax aristotelis   13
Snipe, common: Gallinago gallinago   63
Swallow: Hirundo rustica   61
Swan, mute: Cygnus olor   23. 130
Tern, arctic: Sterna paradisaea   57
Thrush, song: Turdus philomela   iv (nest) 44
Tit, blue: Parus caeruleus   89
  "   coal: Parus ater   110
  "   great: Parus major   124
Treecreeper: Certhia familiaris   63
Turnstone: Arenaria interpres   10
Twite: Acanthis flavirostris   52
Warbler, willow: Phylloscopus trochilus   50
  "   wood: Phylloscopus sibilatrix   62
Wheatear: Oenanthe oenanthe   70
Woodcock: Scalopax rusticola   21
Wren: Troglodytes troglodytes   18. 54

NATURAL SIZE
EGG OF
GREENFINCH

Fishes

Pipefish, snake ; Entelurus aequoreus 126-7
Salmon, : Salmo salar 12
Shanny : Lypophrys pholis 16
Stickleback, 3 spined : Gasterosteus aculeatus V
Trout : Salmo trutta 86

Insects

Ant, black : Lasius niger 77
Bee, buff-tailed : Bombus terrestris 46.73
 " common carder : " agrorum 97
 " leaf-cutter : Megachile centuncularis 98
 " mining : Andrena sp. 98
 " red-tailed : Bombus lapidarius 73
Beetle, cardinal : Pyrochroa coccinea 142
 " dor : Geotrupes stercorarius 63
 " ladybird : Adalia 2-punctata 76.77
 " sexton : Nicrophorus vespilloides 62
Butterfly, common blue : Polyommatus icarus 76?
 " copper, small : Lycaena phlaeus vi.75
 " fritillary, small pearl-bordered : Boloria selene? 88
 " green hairstreak : Callophrys rubi 57.62
 " meadow brown : Maniola jurtina 96
 " painted lady : Cynthia cardui 93
 " peacock ; Inachis io 39
 " Scotch argus : Erebia aethiops xv
 " speckled wood : Parage aegeria 78
 " tortoiseshell, small : Aglais urticae 54.55
 " white, green veined : Artogeia napi 53
 " white, small : Artogeia rapae 53
Cranefly : Tipula paludosa 81
Damsel fly : Coenagrion puella 100
Dragonfly : Aeshna juncea 82
Froghopper (cuckoo spit) : Philaenus spumarius 91?.44
Grasshopper : Chorthippus brunneus 77
Hover fly : Syrphus ribesii 77.90
Ichneumon fly : Netelia testacea 93
Lacewing fly : Chrysopa carnea 75
Mayfly : Ephemera danica 64
Moth, burnet : Zygaena filipendulae 83
 " chimney sweeper : Odezia atrata 87
 " cinnabar : Tyria jacobaeae 77
 " emperor : Saturnia pavonia 112

Moth, fox : Macrothylacia rubi 122
 " garden tiger : Arctia caja 62.93
 " pale tussock : Dasychira pudibunda 117
Wasp, mason : Odynerus spinipes 93
 " ruby-tail : Chrysis ignita 93

Other Fauna

Anemone, beadlet : Actinia equina 19
 " dahlia : Tealia felina 19
Barnacle, acorn : Balanus balanoides 6.92
 " goose : Lepas anatifera 84
Brittle star : Ophiothrix fragilis 42
Cockle, common : Cardium edule 17
Crab, broadclawed porcelain : Porcellana platycheles 42
 " hermit : Eupagurus bernhardus 130
 " shore : Carcinus maenas 42
Frog, common : Rana temporaria 48
Jelly fish, blue : Aurelia aurita 83.84
 " " lion's mane : Cyanea capillata 83.84
Limpet ; common : Patella vulgata 19
Lizard, common : Lacerta vivipara 66
Lobster, spiny squat : Galathea strigosa 105
Mussel, common : Mytilus edulis 16.92
Oyster, Portuguese : Crassostrea angulata 30
Pelican's foot : Aporrhais pes-pelicani 17
Prawn, Dublin bay : Nephrops norwegicus 85
Sea lemon : Archidoris pseudoargus 127
Sea urchin : Echinus esculentus 16.17
Slug, black : Arion ater 63
Snail, glass : not positively identified 118
 " white-lipped banded : Cepaea hortensis 63
Spider, garden : Araneus diadematus 77.98
 " gossamer : Linyphia triangularis 74
Sunstar : Crossaster papposus 139
Toad, common : Bufo bufo 37.54.140
Topshell, grey : Gibula cineraria 19
Whelk, netted : Nassarius reticulatus 17
Winkle, common : Littorina littorea 17

CARDINAL BEETLE

Trees and Bushes

Beech : Fagus sylvatica — 45
Bilberry : Vaccinium myrtillus — 67
Birch, warty : Betula Pendula — 46.102.134
Blackthorn (sloe) : Prunus spinosa — 11.53.119
Bramble (blackberry) : Rubus fruticosus — 66.93.103.104.119
Broom (Common) : Cytisus scoparius — 63
Crabapple : Malus sylvestris — 67
Fuchsia : Fuchsia magellonica — 75.89.106
Gorse (whin) : Ulex europaeus — 2
Hazel : Corylus avellana — 27.109
Hawthorn, common : Crataegus Monogyna — 109
Holly : Ilex aquifolium — 132
Honeysuckle : Lonicera periclymenum — 91.111
Ivy : Hedera helix — 27.29.132
Juniper : Juniperus communis — 119
Larch, European : Larix decidua — 26.32.50
   " Japanese : Larix kaempferi — 137
Osier, common : Salix viminalis — 46
Pine, Scots : Pinus sylvestris — 134
Raspberry : Rubus idaeus — 119
Rhododendron : Rhododendron ponticum — 69
Rose, various : Rosa sp. — 60.94.119
Rowan : Sorbus aucuparia — 54.119.139
Sallow : Salix caprea — 27
Sitka spruce : Picea sitchensis — 133
Sycamore : Acer pseudoplatanus — 117
Yew : Taxus baccata — 119

Flowers, Ferns, Grasses, Rushes, Sedges and Fungi

Anemone, wood : Anemone nemorosa — 52
Angelica : Angelica sylvestris — 3.103
Aster, sea : Aster tripolium — 81
Avens, water : Geum rivale — 81
Balsam, Himalayan : Impatiens glandulifera — 107
Bindweed, greater : Calystegia silvatica — 71.97.76
Bugle : Ajuga reptans — 77
Buttercup, creeping : Ranunculus repens — 77
Butterwort, common : Pinguicula vulgaris — 78
Campion, red : Silene dioica — 64
   " sea : Silene maritima — 67.77
Carrot, wild : Daucus carota — 96

Celandine, lesser : Ranunculus ficaria — 35.46.55
Clover, red : Trifolium pratense — 67.75
   " white : Trifolium repens — 72
Cocksfoot : Dactylis glomerata — 80
Coltsfoot : Tussilago farfara — 40
Columbine : Aquilegia vulgaris — 76
Comfrey, common : Symphytum officinale — 94
Cornsalad : Valerianella locusta — 67
Cottongrass, common : Eriophorum angustifolium — 78
Cranesbill, dovesfoot : Geranium molle — 67.
   " cut-leaved : " dissectum — 77
Cuckoo flower : Cardamine pratensis — 64
Daisy, common : Bellis perennis — 5. 67
   " oxeye : Leucanthemum vulgare — 80
Dandelion : Taraxacum officinale — 103
Everlasting, mountain : Antennaria dioica — 56
Eyebright : Euphrasia officinalis — 75
Forget-me-not, field : Myosotis arvensis — 67.75
   " " water : " scorpioides — 87
Foxglove : Digitalis purpurea — 73
   " fairy : Erinus alpinus — 63
Foxtail, meadow : Alopecurus pratensis — 80
Gentian, field : Gentianella campestris — 108
Globeflower : Trollius europaeus — 78
Grass of Parnassus : Parnassia palustris — 100
Hardfern : Blechnum spicant — 87
Harebell : Campanula rotundifolia — viii. 96
Hawkweed, mouse-ear : Hieracium pilosella — 75
   " orange : Hieracium aurantiacum — 77
Heartsease (wild pansy) Viola tricolor — ii
Heath, cross leaved : Erica tetralix — 87
Heather, bell : Erica cinerea — 76.96
   " ling : Calluna vulgaris — 67.104
Herb Robert : Geranium Robertianum — 76
Horsetail, marsh : Equisetum palustre — 65
Hyacinth : Endymion nonscriptus — 50.64
Iris, yellow flag : Iris pseudacorus — 71.82.107
Knapweed, black : Centaurea nigra — 108
Knotweed, Himalayan : Polygonum polystachium — 100
Lady's bedstraw : Galium verum — 77
Lily, Pyrenian : Lilium pyrenaicum — 81
Lousewort : Pedicularis sylvatica — 81

LOOPER

Marigold, marsh : Caltha palustris   64
Mayweed, scentless : Matricaria perforata   108
Meadow grass, annual : Poa annua   80
Meadowsweet : Filipendula ulmaria   94
Milkwort, common : Polygala vulgaris   67
Montbretia (wild gladiolus) Crocosmia   108
Orchid, broad leaved marsh : Dactylorhiza majalis   62
  " early purple : Orchis mascula   61
  " greater butterfly : Platanthera chlorantha   62
  " heath spotted : Dactylorhiza maculata   62
Pearlwort, knotted : Sagina nodosa   75
Pignut : Conopodium majus   61
Plantain, greater : Plantago major   67
  " sea : Plantago maritima   76
Poppy, Welsh : Meconopsis cambrica   80
Primrose : Primula vulgaris   5. 49
Purslane, pink : Montia sibirica   80
Ragged robin : Lychnis flos-cuculi   87
Ragwort : Senecio jacobaea   102
Ramsons : Allium ursinum   53
Sage, wood : Teucrium scorodonia   78
St. John's wort, slender : Hypericum pulchrum   75
Scabious, devil's bit : Succisa pratensis   108
Scurvygrass, common : Cochlearia officinalis   39, 46
Self-heal : Prunella vulgaris   76
Silverweed : Potentilla anserina   61
Skullcap : Scutellaria galericulata   81
Snowdrop : Galanthus nivalis   vii
Sorrel, common : Rumex acetosa   75
  " wood : Oxalis acetosella   50
Spearwort, lesser : Ranunculus flammula   46
Speedwell, common field : Veronica persica   77
  " germander : " chamaedrys   67, 80
  " grey field : " polita   75
  " water : Veronica anagallis aquatica   87
Spleenwort, maidenhair : Asplenium trichomanes   9
  " sea : Asplenium marinum   9
Stitchwort, greater : Stellaria holostea   64
Stonecrop, English : Sedum anglicum   9. 76
Strawberry, wild : Fragaria vesca   64. 96
Thistle, creeping : Cirsium arvense   87
  " spear : " vulgare   75. 93

Thrift : Armeria maritima   53
Thyme, wild : Thymus serpyllum   11. 77
Toadflax, ivy-leaved : Cymbalaria muralis   57
Toadstool, Clitocybe sp.   87
  " , fly agaric : Amanita muscaria   115
  " Hygrocybe sp.   114
  " lawyer's wig : Coprinus comatus   115
  " sickener : Russula emetica   87
  " sulphur tuft : Hypholoma fasciculare   113
  " various   90
Tormentil : Potentilla erecta   67
Trefoil, birdsfoot : Lotus corniculatus   67, 76, 107
Tutsan, Hypericum androsaemum   119
Valerian, red : Centranthus ruber   76
Vetch, bush : Vicia sepium   66. 67. 107
  " kidney : Anthyllis vulneraria   67
  " tufted : Vicia cracca   83
Vetchling, meadow : Lathyrus pratensis   77
Violet, common dog : Viola riviniana   31. 46. 49. 88
Water lily, white : Nymphaea alba   86
Willowherb, broad leaved : Epilobium montanum   75
  " " rosebay : Epilobium angustifolium   94
Woodrush, great : Luzula sylvatica   65
Yarrow : Achillea millefolium   77

Seaweeds

Coralline : Corallina officinalis   19
Furbelow : Saccorhiza sp.   17
Oarweed : Laminaria saccharina   17
Thongweed : Himanthalia elongata   16
Wrack, knotted : Ascophyllum nodosum   7
  " serrated : Fucus serratus   6

Composite Plates

Summer on Seil   80
(Poppy, Speedwell, Purslane, Foxtail, meadowgrass, Plantain, Oxeye, Cocksfoot)
Autumn on Teesdale   108
(Ling, Montbretia, Mayweed, Bramble, Scabious, Knapweed, Field Gentian)

FROG HOPPER
(Adult)